Sustainable Electricity II

Jessica Fox • Morgan Scott
Editors

Sustainable Electricity II

A Conversation on Tradeoffs

Springer

Editors
Jessica Fox
Electric Power Research Institute
Palo Alto, CA, USA

Morgan Scott
Electric Power Research Institute
Washington, DC, USA

ISBN 978-3-030-07084-7 ISBN 978-3-319-95696-1 (eBook)
https://doi.org/10.1007/978-3-319-95696-1

Softcover re-print of the Hardcover 1st edition 2019

This Springer imprint is published by the registered company Springer Nature Switzerland AG
The registered company address is: Gewerbestrasse 11, 6330 Cham, Switzerland

Foreword

Energy is the key driver of sustainable development, without which achievement of the United Nations Sustainable Development Goals will fall short. Affordable, modern energy is necessary to meet the growing demands of a global society, the pursuit of which must balance economic needs with environmental goals. These trade-offs are no more apparent than in the global goal of providing universal access to modern energy by 2030. As we think about how to generate, distribute, and utilize this energy most effectively and efficiently, we have a collective opportunity to empower global development through electrification.

Even in some of the most progressive energy outlooks, global demand for electricity is expected to increase by over 50 percent between now and 2040.[1] This is driven primarily by developing economies' population growth and increases to economic output. As the world's appetite grows for more modern conveniences like smart appliances, installed cooling, connectivity, and electric transport, governments are working to balance these rising demands with environmental concerns, including improving air quality and reducing carbon emissions.

According to the International Energy Agency (IEA), electricity is becoming the energy of choice in many end-uses. Practically, this means investing trillions of dollars in the coming decades to build generating capacity and distribution infrastructure capable of delivering electricity to current customers and the more than 1 billion who do not have access. Additionally, there are nearly 3 billion without access to improved sanitation. Bridging that divide, which is essential to bringing all people into the modern era, will require significant amounts of new energy.

Today, we reflect on the fact that two milestone agreements, one signed in New York and the other in Paris, are aimed at creating a more sustainable energy future. The Paris Agreement serves as a road map for policymakers for limiting carbon emissions. The UN Sustainable Development Goals establish a universal paradigm for achieving a healthy, poverty-free, prosperous planet. Moving forward, we must grapple with some very tough challenges as we look for innovative, economically feasible, and environmentally responsible solutions to electrify a planet that will soon host over 9 billion people.

[1] IEA 2017 World Energy Outlook

To that end, there are several areas of opportunity worth noting. While not by any means exhaustive, taken together, they can propel ideas forward and help ensure that the overriding objectives of each agreement are met in a balanced and equitable manner:

1. Decoupling GDP growth from electricity demand through improved efficiency
2. Focusing locally to bridge the energy divide in developing countries
3. Leveraging the innovation of business

In North America, despite continued growth in GDP, we are likely to see slower growth for electricity demand in the coming decades.[2] This is due to a number of factors, including significant improvements in energy efficiency and the transition from a manufacturing economy to one that is more service oriented. This decoupling plays an important role in meeting carbon reduction goals in developed nations, as standards of living can be maintained or improved while reducing greenhouse gas emissions. While renewable sources have provided reductions in carbon emissions, the most significant shifts have come from the use of natural gas in the power sector.

Yet this decoupling is more problematic in developing countries, where the establishment of heavy manufacturing and large-scale thermoelectric power generation is the key to economic growth. In this environment, it will be challenging to achieve sustainable energy growth. Also important is the opportunity to find local solutions to local problems. Examples of ways to achieve this goal include adoption of new technology such as large-scale deployment of modern cookstoves and the switch from coal to gas for electricity generation, which can provide some of the most significant improvements in local air quality and related health impacts.

The ability of business to progress the energy system is as powerful today as it was at the turn of the last century, when the industrial revolution brought substantial economic growth and prosperity in the Western world. While we have historically looked to governments to solve some of our most intractable problems, today's solutions will involve considerable contributions from the private sector. We need to look no further than technology companies who are changing not only the delivery of electricity but the generation mix itself. And the oil and gas companies of yesterday are taking steps to provide the energy of the future, a future where oil and gas will remain an important part of the energy equation.

At Chevron, we are working to provide the energy necessary for that future, primarily though the production of natural gas and investments in technology and scientific innovation that will push the frontiers of our industry. As a large user of energy, we are continually examining the trade-offs as we work to improve our energy efficiency and reduce our environmental footprint.

This book, *Sustainable Electricity II: A Conversation on Tradeoffs*, comes at a critical time in the energy policy discussion. Capturing a myriad of topics, including case studies from the electric power sector and perspectives from investors and

[2] IEA 2017 World Energy Outlook.

customer expectations, the book emphasizes some of the most challenging trade-offs in meeting economic, social, and environmental goals. The experiences and perspectives detailed in this book are important to understand as we collectively work toward a sustainable energy future.

The complexity of our global, shared challenge to electrify the world without undue compromise to the integrity of our planet is one that will require global action. It will also require a shared understanding and responsibility between business, government, civil society, and communities to lift billions out of energy poverty. We must be willing to find compromise to allow humanity to thrive, balancing the local with the global and finding new ways of driving economic growth while environments flourish. This is the call to action. The 1 billion plus are counting on us.

Kirsten S. Thorne
Global Issues and Public Policy Manager, Chevron

For nearly two decades, Thorne has been working to advance environmental and social policy at Chevron, one of the world's leading integrated energy companies. She has worked across many facets of sustainability, including environmental and social performance and corporate reporting. She has a JD from Chapman University School of Law with an emphasis in land use and environmental policy.

Generation Z on Sustainable Electricity

As we explore the meaning of "sustainable electricity" and how best to protect resources for future generations, we discover that as a society we often leave members of the future generation out of the conversation altogether. Generation Z (Gen Z), also known as the post-Millennials, entered the world as early as the second half of the 1990s and has already begun to enter the workforce.[1] As the next generation to inherit this planet, they deserve a seat at the table to contribute to this dialogue. As such, the first volume of this book, *Sustainable Electricity: Case Studies from Electric Power Companies in North America,* provided perspectives from a variety of Gen Zs. For this book, we are pleased to present the winning essay submission from EPRI's employee-driven Corporate Social Responsibility Essay and Art contest. The following winning submission provides insight into considerations around power generation and the impacts on the environment. It provides a glimpse into the expectations of a generation that was raised with a recycling bin in every classroom, has a sophisticated understanding of environmental issues, and experiences increasing reliance on electricity for even the most basic daily tasks. While this is just one perspective, it is shared with other Gen Zs and could very well redefine an entire industry, and our world.

– Jessica Fox and Morgan Scott

Andrew Voelsing, 13 years old

In our modern society we have found ourselves relying on electricity for nearly everything we do. Being a stereotypical kid these days, I rely on electricity for my comfort, my entertainment, and my education. Electricity powers all my personal electronics, the computers at my home and school, as well as the lights and air-conditioning everywhere I go. There are many ways to make electricity, but different processes using different types of fuel can have different impacts on our planet. Maybe we should look toward more environmentally friendly ways to supply our lives with electricity. Our forms of energy production are good for now, but we need

[1] Pew Research, 2018 (http://www.pewresearch.org/fact-tank/2018/03/01/defining-generations-where-millennials-end-and-post-millennials-begin/)

energy that once produced can be reproduced. Modern technology is constantly evolving in ways that increase our reliance on electricity. If our way of generating electricity does not evolve to be sustainable, then we will deplete the resources and deny future generations. That is why I think we, as people who use electricity, should strive to create better forms of sustainable electricity.

Today we produce electricity in many different ways. The current large electrical generators create power by burning fossil fuels, such as coal or natural gas, or by using a nuclear reaction, known as fission. We use the methods because they are easy ways to mass generate electricity, right? So what harm could they do? Well, more than you might think.

The USA has been on a decline in coal power consumption in recent years, but why? One reason is there can be significant environmental impact associated with coal power generation. When the fuel (coal) is burned, it produces emissions which pollute the air. One example is carbon dioxide (CO_2) emissions, which impact global warming. Coal generation also produces sulfur dioxide (SO_2) and mercury emissions, two other forms of air pollution that impact air quality and can lead to things like acid rain. One example I read about in school was how the air quality and visibility in the Great Smoky Mountains, near my home in North Carolina, improved after nearby coal-burning power plants were shut down. This is a great example of why we need to consider other sources of electricity. First of all, our supply of coal is not infinite. Also, burning coal hurts the environment, which impacts not just our planet but the quality of our lives.

Another form of power generation is nuclear power. The USA receives nearly one fifth of our power from nuclear power plants, but how does it work? Nuclear power plants use the energy generated by splitting the atoms of certain elements to produce electricity. This provides us with an inexpensive alternative to burning fossil fuel that does not generate carbon emissions. It sounds good, but there are some downsides. Nuclear power plants produce radioactive isotopes that can be dangerous if they come into contact with humans or the environment. Natural radiation exists all around us, but the levels associated with nuclear power are much higher and can be harmful. The nuclear fission reaction produces radioactive materials that must be contained and properly disposed of to prevent radiation exposure or contamination. It must be stored safely until it no longer emits a dangerous level of radiation, which can take up to ten thousand years. Once again, the challenges associated with fuel for nuclear raise concerns that we should consider when deciding how we generate power.

Now we are looking for environmentally friendly ways to produce electricity. To me, "sustainable electricity" is electricity produced in a way that does not harm our environment and comes from a fuel source that can be replenished. We already have some ways to do that – such as wind turbines which harness wind power, solar panels which harness the sun, and hydroelectric plants which harness water from a river or reservoir. The problem is these methods cannot produce enough electricity to supply our lives and they cannot produce electricity at all times. It looks like these problems may be left for my generation to fix. I believe research can help us to solve these problems. We can look at ways to use less power and to store electricity to use when the wind is not blowing or the sun is not shining. We can also work on more radical research such as totally new ways to generate electricity. Personally, I am really interested in the possibility of nuclear fusion, which can generate huge amounts of electricity with tiny amounts of fuel and few environmental impacts.

The environment is important to me, because it is the place that gives me all that I could ever need. If I, and all other humans, do not take care of it today, then we are setting up a major issue for people to come. As I see it, this planet gives everything to me so in return I must nurture and care for it as a sort of mutual relationship. The more I give to it, the more it gives back to all of us. Part of protecting the planet is thinking about the environmental impacts of the power we generate and use in our everyday lives. I have thought about this and I have realized we have only scratched the surface of electricity, and if we look toward the future and truly focus on protecting the planet, we can one day change electricity as we know it.

Preface

Societies and industry are struggling with managing the unprecedented transformation of the electric power system that will continue throughout the next decade. Responding to evolving customer interest in their energy choices, aging infrastructure replacement, physical asset retirement, and cybersecurity threats while simultaneously managing for sustainability can be daunting. Building on the first volume, *Sustainability Electricity: Case Studies from Electric Power Companies in North America* (Fox, 2016, Springer), this book dives into the economic, environmental, and social tradeoffs power companies face in making difficult sustainability choices. Readers are given a window into how electric power companies and their stakeholders balance conflicting priorities to address sustainability challenges while meeting the industry's core mandate of providing safe, reliable, affordable, and environmentally responsible electricity. Uniquely, a variety of stakeholder perspectives are included from customers, regulators, investors, and environmental groups. With more than 40 expert authors and interviewees, the book compiles a truly unique resource. As demonstrated in the chapters, not everyone will agree on the proper balance of tradeoffs: greenhouse gas emissions reductions vs. affordable power, species habitat vs. renewable energy, transparency vs. disclosure cost burden, and current reliability vs. long-term resiliency. Surprisingly, however, some of the toughest tradeoffs can be avoided through creativity and boldness. This book provides a foundation needed to advance a larger conversation around the meaning of sustainable electricity in a time of significant industry transformation.

Palo Alto, CA, USA　　Jessica Fox
Washington, DC, USA　　Morgan Scott

Acknowledgments

We would like to thank all the contributing authors, who were subject to rounds of demanding edits and questions. All authors worked hard to ensure the quality, candidness, and insight of their contributions. We also thank the members in the Electric Power Research Institute's Energy Sustainability Interest Group for making this book possible and creating an important strategic resource that will advance a broader conversation on sustainability tradeoffs in the electric power industry.

Last but not least, we thank Ellen K. Donnelly of Donnelly Communications for her tireless support in the copyediting and compilation of the book, yet another example of her dedication to EPRI's sustainability research for over 10 years.

Contents

About the Editors

Jessica Fox Fox is a Senior Technical Executive at the Electric Power Research Institute (EPRI), where she has managed EPRI's $15 million Water & Ecosystems research area and is an expert in environmental markets and "credit stacking," ecosystem services, and corporate sustainability. Fox created the EPRI Ohio River Basin Water Quality Trading Project, which is the world's first interstate trading program for nutrients (http://wqt.epri.com), and won the US Water Prize in 2015. Fox led EPRI's Energy Sustainability Interest Group from 2011 through 2016 to become the largest collaborative forum in the industry to strategically advance critical issues related to the sustainable generation and distribution of electricity, including consensus by 45 utilities on sustainability metrics to use to benchmark their companies' performance (www.epri.com/sustainability). Fox launched EPRI's internal corporate social responsibility program in 2016 under EPRI's Chief Sustainability Officer and CEO. In 2017, Ms. Fox was recognized as Top Innovator (The Fantastic Forty) by *Public Utilities Fortnightly*. In 2018, Ms. Fox launched the Power-in-Pollinators Initiative, which is set to be the largest effort in North America focused specifically on pollinator conservation at electric power companies. Her projects have been covered by the *Wall Street Journal*, *The Economist*, National Public Radio, Bloomberg, and others. She is a trained meeting facilitator and conflict resolution mediator, a member of the Ecological Society of America, and a Certified Habitat Steward by the National Wildlife Federation. Prior to EPRI, she worked as a molecular biologist leading conservation genetics projects on dragonflies in Costa Rica, at Stanford University sequencing and mapping the human genome, and as a manager at a biotech company developing DNA analysis technologies. She has a Bachelor of Science degree from the University of California, Davis, and a Master of Science degree from Stanford University.

Morgan Scott Scott is a Senior Sustainability Technical Lead at the Electric Power Research Institute (EPRI). In this role, Scott leads EPRI's growing portfolio of sustainability research, focused on developing the tools and resources electric power companies can use to establish and enhance their sustainability programs as well as embed a triple bottom line mindset into their long-range planning. Scott manages

EPRI's Strategic Sustainability Science program, Energy Sustainability Interest Group, and Sustainability Benchmarking for Utilities project in collaboration with over 45 electric utilities throughout North America. Prior to joining EPRI, Scott held positions of increasing responsibility at Consolidated Edison Company of New York, Inc., culminating with her role as their Sustainability Manager responsible for the company's sustainability strategy and associated initiatives. Scott received a Bachelor of Science degree in Business Administration from Wagner College and a Master of Science degree in Sustainability Management from Columbia University.

Chapter 1
Introduction: The Tradeoff Conversation for Sustainable Electricity

Jessica Fox and Morgan Scott

Abstract Societies and industry are struggling with managing the unprecedented transformation of the electric power system that will be occurring over the next decade. This book focuses on how companies and stakeholders are balancing the choices and tradeoffs with electricity generation and use. Uniquely, a variety of perspectives are included from power companies, customers, regulators, investors, and environmental groups. Not everyone will agree on the proper balance of tradeoffs. This book provides a foundation needed to advance a larger conversation around the meaning of sustainable electricity in a time of significant industry transformation.

What is a tradeoff? Academically, a *tradeoff* is an exchange in which something desirable is given up for another thing regarded as more desirable; a balance achieved between two desirable but incompatible elements [1]. Practically, we assess tradeoffs based on an analysis of the pros and cons of different choices. What are we giving up in making one choice over another? What compromises, particularly when they're made with stakeholders, are tolerable? When the compromises are deemed acceptable, even when difficult, the tradeoff is made.

The less heterogeneous and the more limited the variables, the easier the analysis. For example, if your only concern is spending money, you will purchase the cheapest product. For illustrative purposes, let's consider the purchase of a new sofa. If you have a limited budget, you may select the cheapest sofa you can find that will fit in your room. If you are concerned about the longevity of the sofa, you may be willing to pay more for something that will last longer. If you are also worried about fair labor practices and the sourcing of the materials, you might add criteria for the sofa to be made locally with domestic materials. Maybe you will buy a sofa only from a store that demonstrates a broader commitment to sustainability. With the additional criteria, your sofa options, and the evaluation of those options, become more complicated. Maybe the sofa that is higher quality doesn't come in the

J. Fox (✉) · M. Scott (✉)
Electric Power Research Institute, Palo Alto, CA, USA
e-mail: jfox@epri.com; mmscott@epri.com

J. Fox, M. Scott (eds.), *Sustainable Electricity II*,
https://doi.org/10.1007/978-3-319-95696-1_1

Fig. 1.1 Balancing sustainability tradeoffs (Drawing provided by http://www.nityawakhlu.com/) [2]

color you want. The sofa made in the United States may be twice as expensive, but it is the style you want. Durability vs. cost, environmental footprint vs. ascetics. Which sofa do you buy? The more criteria and variables, the more complicated the decision. Is there any way to have it all? (Fig. 1.1).

Relatively simple tradeoff analysis can be done when the variables are measured using the same units. For example, if all you care about is money, you will want to evaluate all variables in consideration of economic units and impact (e.g., dollars), which can be more challenging for considering things like environmental impacts and opportunities (e.g., ecosystem services). Tradeoffs between sustainability pillars,[1] for example, economic vs. environmental or social vs. environmental considerations, are more complicated partly because they require analysis measured in different units: acres of habitat, dollars, tons of greenhouse gases, number of power outages, human safety incidents, or volunteer hours. Advancement toward clean electricity has been challenged by the complexity of identifying specific actions that appropriately balance economic, environmental, and social targets. This is indeed a complex consideration, the solution for which may change depending on time and place variables.

In further contemplating the complexity of the triple bottom line,[2] Fox previously noted,

> It seems that the [triple bottom line] concept doesn't adequately recognize the three timelines of when the bottom line is reconciled. Financially, companies will roll-up analysis annually; but more likely quarterly as they continually track economic progress, profits and loss, spending, and revenues. In contrast, ecologists think in terms of decades, if not centu-

[1] The three sustainability pillars are economic, environmental, and social responsibility.

[2] Triple bottom line generally refers to simultaneously positive outcomes across environmental, economic, and social components, particularly resulting from corporate decision-making.

> ries, for the environmental bottom line. For them, reconciliation occurs every 10 years, give or take 5-100 years depending on the natural resource. The social bottom line is arguably somewhere in between but certainly will occur several times within the lifetime of a single person (maybe every 5-10 years). With reconciliation happening at different intervals from once a quarter to every 100 years, this differential possibly reveals a gap between the triple bottom line *concept* and the reality of its boardroom *application*. While the concept has motivated very important action, the sophistication of the idea and the resulting corporate examples needs to advance [2].

Identifying a specific approach for achieving sustainable electricity is further stymied by the large diversity of power companies. The electric power industry is comprised of thousands of companies, some that are owned and operated by various government entities, others operating as profitable enterprises, and others functioning as customer-owned cooperatives. Today, the influx of nontraditional power providers, particularly in solar and wind generation, adds more diversity. There may also be differences between a sustainable kilowatt hour and a sustainable electric power company (i.e., commodity vs. corporate sustainability); conversations around "sustainable electricity" may need to clarify which issue is being discussed. As such, it is not possible to have a single step-by-step guide that identifies the approach within one company, let alone the whole industry.

Sustainable Electricity: Case Studies from Electric Power Companies in North America [3] was the first compilation of industry-told case studies on how power companies are making strides toward achieving sustainable electricity. Chapters discussed hotly debated challenges: renewable energy, water use, species impacts, employee engagement, stakeholder communication, resiliency and climate change, distributed energy, consumer preferences, and business vitality. Also discussed was the meaning of "sustainable electricity" as evolving and needing to include specific goals and targets that will "protect future generations" [4].

In developing this new book, *Sustainable Electricity II: A Conversation on Tradeoffs*, we wanted to advance a deeper discussion of the choices related to both sustainable electricity and a sustainable electric power company. We dive into specific tradeoff scenarios to understand in detail how choices are made by companies. We include first-person perspectives from environmental groups, investors, customers, and regulators to broaden the conversation. We were motivated by a festering frustration at hearing only corporate stories that result in triple bottom line wins, rather than discussing tougher choices that often face the heavily managed and regulated electric power industry. Further, we are challenged by the complexity of tradeoffs and the creation of value, making decisions when certain choices support a sustainable economy at a high level, but resulting in tradeoffs for individual parties. Electrification is one example of this type of value tradeoff consideration. In the pursuit of reduced greenhouse gas emissions, electrification of technology in sectors like transportation could result in significant regional emission reductions [5]. However, this could potentially mean level or a slight initial increase of emissions at the individual electric power company level, dependent on its fuel source profile. Is this type of tradeoff acceptable?

Specifically, in this book we wanted to further consider the following:

> While environmental and social benefits can be co-benefits of financial outcomes in the electric power industry, we need to better understand WHICH decisions provide these

co-benefits. What are the choices that lead to those agreeable outcomes? For the electric power industry, we need to know the "right" decisions when win-win-win outcomes are not clear [6].

As we see throughout the chapters in this book, not all tradeoffs can be easily resolved. For example, American Electric Power, one of the largest electric utilities in the United States, describes very difficult choices for meeting conflicting environmental and economic goals in reclaiming environmental, social, and economic value from former mine lands (Chap. 3). Los Angeles Department of Water and Power, the largest municipal utility in the United States, highlights the importance of equity as the ubiquitous core value that drives their commitment to transparency and customer engagement but also necessitates some tradeoffs (Chap. 7). Ameren considers the various choices, values, and costs associated with their annual Corporate Social Responsibility Report publication (Chap. 2). These chapters illuminate how companies balance difficult and often conflicting choices.

The book also adds several new examples of when creativity, innovation, and flexibility lead to solutions that minimize the toughest tradeoffs and achieve, or come close to, triple bottom line success. Portland General Electric shows how creativity in regulation significantly reduced the cost impacts of emission reductions, improved flexibility in meeting carbon reduction goals, created a community solar program, and allowed utilities to invest in transportation electrification using customer funds. This collaborative effort limited the total possible tradeoffs involved in reducing carbon emissions while controlling customer costs (Chap. 8). Santee Cooper led a creative effort to dispose of coal ash through recycling, resolving the hardest compromises between the company and community (Chap. 10). Exelon, serving the largest number of customers of any power company in the United States, shows how they embrace innovation as a means to respond to changing technology and evolving customer demands (Chap. 5).

Uniquely, this book does not shy away from diverse views. A former regulator discusses the drivers both influencing and influenced by the utility regulatory model and explores the tools available to regulators to enable cost-effective grid modernization while balancing the affordability needs of the consumer and the financial stability of utilities (Chap. 6). A former regulator and lifelong environmentalist questions whether tradeoffs are really needed in energy climate policy (Chap. 12). A socially responsible investor reveals that financial performance and sustainability are compatible in many instances, but is complicated for the power sector as it manages issues related to emissions, waste generation, and industry transformation (Chap. 4). Going deeper, a sustainability expert in the Silicon Valley captures the expectations for clean energy solutions from some of the largest and most innovative electricity customers in the world, including Facebook, Google, VMware, Hewlett Packard, and Stanford University (Chap. 9).

Societies and industry are struggling with managing the unprecedented transformation of the electric power system that will be occurring over the next decade. Responding to renewable power requirements, aging infrastructure replacement, physical asset retirement, and cybersecurity threats while simultaneously managing for sustainability can be daunting. With more than 40 expert authors and interviewees, this book compiles a truly unique resource and provides a foundation needed to

Fig. 1.2 EPRI Energy Sustainability Interest Group members 2018

advance a larger conversation around the meaning of sustainable electricity in a time of significant industry transformation.

This book was sponsored by EPRI's Energy Sustainability Interest Group, which is the largest collaboration in the electric power industry focused specifically on sustainability issues. The group was launched by EPRI in 2008 and consistently has over 40 members with combined assets of over $1.1 trillion. The focus of the group is to help electric power companies establish and enhance their sustainability programs through collaboration, technical research, and opportunities to interact with key sustainability players (Fig. 1.2).[3]

Author Profiles

Jessica Fox Fox is a Senior Technical Executive at the Electric Power Research Institute (EPRI), where she has managed EPRI's $15 million Water & Ecosystems research area and is an expert in environmental markets and "credit stacking," ecosystem services, and corporate sustainability. Fox created the EPRI Ohio River Basin Water Quality Trading Project, which is the world's first interstate trading program for nutrients (http://wqt.epri.com), and won the US Water Prize in 2015. Fox led EPRI's Energy Sustainability Interest Group from 2011 through 2016 to become the largest collaborative forum in the industry to strategically advance critical issues related to the sustainable generation and distribution of electricity, including consensus by 45 utilities on sustainability metrics to use to benchmark their companies' performance (www.epri.com/sustainability). Fox launched EPRI's internal corporate social responsibility program in 2016 under EPRI's Chief Sustainability Officer and CEO. In 2017, Ms. Fox was recognized as Top Innovator

[3] Information on EPRI's Energy Sustainability Interest Group can be found at www.epri.com/sustainability

(The Fantastic Forty) by *Public Utilities Fortnightly*. In 2018, Ms. Fox launched the Power-in-Pollinators Initiative, which is set to be the largest effort in North America focused specifically on pollinator conservation at electric power companies. Her projects have been covered by the *Wall Street Journal*, *The Economist*, National Public Radio, Bloomberg, and others. She is a trained meeting facilitator and conflict resolution mediator, a member of the Ecological Society of America, and a Certified Habitat Steward by the National Wildlife Federation. Prior to EPRI, she worked as a molecular biologist leading conservation genetics projects on dragonflies in Costa Rica, at Stanford University sequencing and mapping the human genome, and as a manager at a biotech company developing DNA analysis technologies. She has a Bachelor of Science degree from the University of California, Davis, and a Master of Science degree from Stanford University.

Morgan Scott, EPRI Scott is a senior sustainability technical lead at the Electric Power Research Institute (EPRI). In this role, Scott leads EPRI's growing portfolio of sustainability research, focused on developing the tools and resources electric power companies can use to establish and enhance their sustainability programs as well as embed a triple bottom line mindset into their long-range planning. Scott manages EPRI's Strategic Sustainability Science program, Energy Sustainability Interest Group, and Sustainability Benchmarking for Utilities project in collaboration with over 45 electric utilities throughout North America. Prior to joining EPRI, Scott held positions of increasing responsibility at Consolidated Edison Company of New York, culminating with her role as their sustainability manager responsible for the company's sustainability strategy and associated initiatives. Scott received a Bachelor of Science degree in Business Administration from Wagner College and a Master of Science degree in Sustainability Management from Columbia University.

Company Profile

Electric Power Research Institute The Electric Power Research Institute, Inc. conducts research and development relating to the generation, delivery, and use of electricity for the benefit of the public. An independent, nonprofit organization, EPRI, brings together its scientists and engineers as well as experts from academia and industry to help address challenges in electricity, including reliability, efficiency, affordability, health, safety, and the environment. EPRI also provides technology, policy, and economic analyses to drive long-range research and development planning and supports research in emerging technologies. EPRI members represent 90% of the electric utility revenue in the United States with international participation in 35 countries. EPRI's principal offices and laboratories are located in Palo Alto, Calif.; Charlotte, N.C.; Knoxville, Tenn.; and Lenox, Mass.

References

1. *The American Heritage Dictionary*. Third Edition. 1992. Houghton Mifflin Company.
2. J. Fox, "The Next Decade of Sustainability Science." In: Fox, J., Editor, *Sustainable Electricity*. (Book) May 2016. Springer Press.
3. J. Fox, Editor, *Sustainable Electricity*. (Book) May 2016. Springer Press.
4. J. Fox and A. Ray, "Defining Sustainable Electricity," In: Fox, J., Editor, *Sustainable Electricity*. (Book) May 2016. Springer Press.
5. U.S. National Electrification Assessment. EPRI, Palo Alto, CA: 2018. 3002013582.
6. J. Fox, "The Next Decade of Sustainability Science." In: Fox, J., Editor, *Sustainable Electricity*. (Book) May 2016. Springer Press.

Chapter 2
Achieving Balance: Ameren's Corporate Social Responsibility Reporting Journey

Ajay Arora, Brad Brown, Diane Saftic, Richard Smith, and Steven Whitworth

Abstract Interest in energy sustainability has grown exponentially in the past 20 years. Energy providers have always had to carefully consider the trade-offs of their near- and long-term decisions. Now, however, the need for greater transparency concerning associated trade-offs has driven energy companies to not only tackle these issues in their long-term planning but also to engage stakeholders along the way. As such, corporate social responsibility (CSR) reporting has become an important communication tool to engage varied stakeholders. While producing transparent and meaningful CSR communications can be a daunting task, a well-constructed CSR report can serve as a useful window into an industry filled with complex decision-making. Our findings and best practices are easily transferable to other companies seeking to create CSR reports for the first time or to optimize existing sustainability communications.

2.1 Introduction

Ameren has historically participated in environmental reporting, integrated resource planning, and financial reporting. As stakeholder interest in the energy sector grew over time, Ameren recognized the need to engage our stakeholders in a transparent dialogue regarding our business perspective on sustainability. Leadership and many others inside the organization knew that our message about the company's efforts regarding sustainability was getting lost among all the information available about our operations, both on our website and in regulatory filings.

Ameren needed stakeholders to understand that we were planning and acting for the future. In particular, our company wanted to articulate how Ameren balances the "triple bottom line" of people, planet, and profit. At the same time, Ameren's various stakeholders were seeking more information about company plans regarding future investment and risks related to coal-based generation—including progress

---

A. Arora · B. Brown · D. Saftic (✉) · R. Smith · S. Whitworth
Ameren Corporation, St. Louis, MO, USA
e-mail: DSaftic@ameren.com

J. Fox, M. Scott (eds.), *Sustainable Electricity II*,
https://doi.org/10.1007/978-3-319-95696-1_2

toward achieving specific goals to reduce air emissions, commodity risks (e.g., cost of fuels), construction risks, and the costs of environmental compliance. We sought to share this view with all stakeholders, including shareholders, customers, elected officials, regulators, and other policymakers; community, business, and civic leaders; nongovernmental organizations; our workforce; and the media.

Communicating to each of these disparate stakeholder groups was a challenge. In many ways, the sustainability approach we articulated at the beginning of our CSR journey echoes the focus of this book: how do energy companies determine priorities, reckon with trade-offs, and develop solutions that are beneficial to its multiple stakeholders? Ameren is part of an industry that has always required thoughtful planning. The reliable service that our customers enjoy today is the result of diligent planning that occurred decades ago. The same is true as we look toward the future. To provide safe, reliable, and reasonably priced energy, Ameren—like many of our industry peers—will continue to work hard to strike the appropriate balance. This is particularly true as new energy technologies evolve, affecting our operations and customers' expectations. Just as our business evolves, so too has our sustainability reporting. With each subsequent report, we have learned new lessons to improve the next publication, continuing to strive to transparently disclose the information our stakeholders are interested in exploring.

Ameren's overall approach to sustainability can be summarized as "achieving balance." This theme has been consistent throughout our CSR reporting journey, starting with this thought from then-CEO Thomas R. Voss' introduction in our 2011 CSR report:

> We believe that meeting our corporate social responsibility requires a careful integration of people, planet and profit. Achieving that balance will elevate customer and shareholder value [1].

Over the years, Ameren teams have spent considerable time building out CSR content that explains the need to achieve balance in our strategic planning and decision-making. In fact, our 2016 CSR report elevated "Achieving Balance" to the report title, with an entire section being dedicated to our approach.

So, what does achieving balance mean? As an investor-owned energy company, Ameren has a social responsibility in four key areas (Fig. 2.1). When making

Fig. 2.1 Ameren seeks to achieve balance with both short- and long-term plans among four key areas of responsibility

short- or long-term plans, we must weigh and carefully balance multiple priorities to help to make them truly sustainable.

For example, as Ameren transitions to cleaner, renewable, and more diverse generation sources, our goal is to do so in a way that preserves system reliability while helping minimize adverse price impacts on our customers. We must achieve the right balance to ensure customer reliability, affordability, satisfaction, and environmental stewardship. As Ameren seeks to replace aging infrastructure with smart grid technology and pursue innovative customer-focused solutions, we must consider our goal in relation to helping enable and sustain the progressive state and federal regulatory frameworks in which we operate to balance shareholder interests.

We have a responsibility to:

- *The customers and communities we serve*. Energy reliability is the bedrock of our region's economic prosperity and security. Ameren has provided, and seeks to maintain, top-tier reliability. We have made strides in recent years to reduce the length and frequency of outages. At the same time, electricity delivered by Ameren remains reasonably priced. This is especially important to low- and fixed-income customers and to businesses seeking to launch or expand in our service territory. Our rates in Illinois and Missouri are below the Midwest and US averages. Customer affordability is an essential theme that must be built into all sustainability goals and reporting.
- *Our workforce of 8600 people, who are our greatest asset*. Ameren's future success is dependent on a capable, diverse, and innovative team. Our 2.4 million electric and 900,000 natural gas customers count on our employees for safe, reliable, affordable, and cleaner energy. That compels our workforce to remain focused on driving efficiencies into all aspects of our operations and to look for ways to innovate. We want to assure a sustainable supply of qualified coworkers over the long term. It's important to build strong relationships with educational institutions and to devote resources to succession planning and workforce development. Most importantly, to provide a sustainable workforce, Ameren must continue to focus on a values-based, safety culture.
- *The environment we share*. We want to make sure stakeholders understand how we are minimizing environmental impacts. For example, Ameren is investing in transmission infrastructure that provides the flexibility needed to use cleaner sources of generation, such as wind energy. We identify—and publicize—expenditures and measures made to reduce air emissions. We are connected to the natural resources that help power our business, and we must protect those resources for future generations. As responsible stewards, we undertake a range of efforts to minimize how our operations impact land, air, and water. For instance, we are focused on using coal combustion residuals (CCRs) for beneficial uses, such as cement and concrete. We are taking steps now to reduce the amount of water used for managing CCRs as we transition to dry ash handling processes. We are also advancing our commitment to reducing greenhouse gas emissions with plans to significantly increase our renewable energy portfolio and targets to

reduce Ameren Missouri CO_2 emissions by 80% by 2050. In addition, in collaboration with the state of Missouri, we have created habitat areas to protect wildlife and provide public recreation. This includes the Reform Wildlife Management Area on 6000 acres of land surrounding our Callaway Energy Center.
- *Our shareholders.* Shareholders are our owners, and their investments make our business model work. They are entrusting us with their money so that they can sustain their financial goals. Maintaining our vast energy infrastructure is capital intensive. Without investor dollars, Ameren would not have all the capital needed to undertake vital system upgrades. Ultimately, shareholder investment and resulting profit drive our ability to serve people and protect our planet. We aim to deliver superior value to our shareholders by delivering strong earnings growth and dividends.

We came to understand that the many public disclosures and pieces of data did not, and could not, communicate the much larger message of sustainability and the steps we are taking to achieve the necessary balance between the multiple aspects of our business. As we strove to improve Ameren's sustainability narrative, our CSR reporting efforts evolved over several cycles, providing valuable and unique opportunities to communicate relevant issues more effectively.

This chapter illustrates the collaborative processes Ameren used to develop our CSR reports over time. It starts with a discussion of our overall approach to sustainability which involves carefully calibrating our important—but sometimes competing—responsibilities to our customers and communities, our workforce, the environment, and our shareholders. The narrative traces the evolution of Ameren's voluntary reporting efforts, starting with our first report in 2011 to our award-winning 2016 edition and culminating with our 2017 report.

2.2 Our First CSR Report (2011): Laying the Groundwork

Ameren began its reporting journey in early 2011. The 2011 effort was initiated out of a desire to communicate transparently with a variety of stakeholder groups. We started our reporting journey from scratch, and our goal was to create one report that integrated environmental, social, and economic elements. Our first step was to establish an internal, cross-functional working team to help guide our reporting process (i.e., a CSR team) that included representatives from Corporate Planning, Corporate Communications, Environmental Services, Finance, Legal, Ameren Illinois, and Ameren Missouri. The team's guiding principles were:

1. *Identify and strive to eliminate gaps* in company efforts to support the triple bottom line of people, planet, and profit.
2. *Focus on current corporate initiatives* involving customers, reliability, communities, employees, operations/environment, and financial/governance/investor.
3. *Provide a single report summarizing* Ameren's commitments, plans, successes, and challenges.

4. *Highlight issues* in a thoughtful, meaningful manner; do not produce a marketing piece.

The CSR team began by researching other sustainability reports and developing an outline. We quickly realized we had a lot of content to share, and we were unsure of what to title our document: Sustainability Report? Corporate Accountability Report? Corporate Responsibility Report? This prompted some reflection. The team looked for advice from the Global Reporting Initiative (GRI)—which described a sustainability report as a publication that:

- Describes the "economic, environmental, and social impacts" caused by a company or organization's everyday activities
- Presents the "values and governance model" of the company/organization
- Demonstrates a link between the company or organization's "strategy and its commitment to a sustainable global economy" [2].

GRI highlighted Ameren's need to balance the triple bottom line to effectively meet our social responsibilities. Their approach was not only forward-looking around sustainability impacts but also retrospective around previous commitments. Overall, GRI focused us on the importance of integrating community betterment, environmental stewardship, and financial strength into our reporting.

"At the time, we realized that 'sustainability' meant different things to different stakeholders. Based on our review of GRI and other sustainability reports, as well as several internal discussions, we chose 'Corporate Social Responsibility Report.' To us, it represented our broader assessment of corporate initiatives, impacts on the environment, social and governance factors, and taking responsibility for our actions," said Geralynn Lord, Vice President, Corporate Communications, Ameren Corporation.[1]

Our 2011 CSR report established a foundation that detailed our history along with opportunities and obstacles we faced. This first version provided a transparent view of our thought process as we were dealing with key energy matters. Many internal subject matter experts provided content, which was then organized and edited by the CSR team. After extensive internal reviews, the report was finalized in December 2011. It was printed and sent to various stakeholders via mail. We also posted a PDF to our website, Ameren.com.

2.2.1 Lessons Learned

The 2011 CSR report was very detailed, was text heavy, and ran more than 60 pages (Fig. 2.2). It was produced in a printed format. When the CSR team considered potential improvement opportunities for future reports, we decided the next report should be more dynamic and better organized. We also wanted to make our next report more accessible and more eco-friendly by making it available online.

[1] Personal Communication, March, 2017.

06 | AMEREN | CORPORATE RESPONSIBILITY

FOCUSED ON PEOPLE, PLANET AND PROFIT

At Ameren, we have committed ourselves to meeting our customers' energy needs, while enhancing shareholder value. Our promise to our investors, customers, employees and the communities we serve is that we will continue to deliver safe, reliable energy in an environmentally responsible manner today, while striving to find even better ways to do that in the future.

That is our focus — now and for life — thus, our new promise: Focused Energy. For Life. Corporate social responsibility is clearly part of that focus, that commitment.

Contents

SUPPORTING SUSTAINABILITY – A WORK IN PROGRESS

While Ameren Corporation has reported a significant amount of information in its public filings and in reports focused on the environment and the community, this is our first Corporate Social Responsibility Report. We have supplemented the content found here with a video, which will be available on our website and, by request, in DVD format for mailing. We also provide greater detail on all the topics covered here on our website at Ameren.com.

Since this is the first Corporate Social Responsibility Report for Ameren, we did not strictly follow the Global Reporting Initiative (GRI) or other sustainability reporting guidelines. We did, however, organize this publication into core areas around the triple bottom line: People, the Planet and Profit. And you will discover that the report does generally meet the following partial elements of the "Sustainability Report Content" test as prescribed by the GRI:

- The main topics and future challenges reported by peers and competitors.
- Relevant laws and regulations with strategic significance to the organization and its stakeholders.
- Key organizational values, policies, strategies, operational management systems, goals and targets.
- The interests/expectations of stakeholders specifically invested in the success of the organization (e.g., customers, communities, employees, shareholders and suppliers).
- Significant risks to the organization.
- Critical factors for enabling organizational success.
- The core competencies of the organization and the manner in which they can or could contribute to sustainable development while providing for the needs of the customers and communities we serve.

Fig. 2.2 Ameren's first CSR report. Produced in 2011, it was text heavy

2.2.2 *Trade-off: More Reporting or Not*

As a regulated, investor-owned utility, Ameren already disclosed a plethora of sustainability-related content. A fundamental trade-off that we had to consider was whether or not to continue to invest in a focused sustainability report. Information about our governance, business model, environmental stewardship, risks, and other subjects was already available through Securities and Exchange Commission filings, regulatory filings, our website, and other venues. That information, however, wasn't neatly summarized or presented in the context of sustainability. Ameren faced a decision on whether to continue creating a separate CSR report.

Deciding to commit to voluntary CSR reporting was not easy. Our first report was resource intensive and took about a year to develop. After publication, Ameren leadership decided we should strive to prepare future CSR reports every other year, reducing the resource burden but still allowing for continued sustainability disclosure. The trade-off was worth it. The decision to continue voluntary CSR reporting provided a pathway to engage with our stakeholders and environmental groups on some of the key issues and gaps we identified. We engaged in several meetings with these stakeholders after the first report was released to further explore energy issues, and this dialogue, in turn, led to improvements to our approach and the quality of information provided.

2.3 Our Second CSR Report (2013): Streamlining Our Focus

In 2013, our company was changing. Ameren was finalizing an exit from our merchant (unregulated) generation business. To help with the development of our second CSR publication, we solicited feedback on energy matters from our customers and stakeholders via surveys, letters, meetings, social media, and other communications channels. Indeed, "engagement" was a new content area for 2013. Following best reporting practices, Ameren explained how customer and stakeholder input informs our decisions.

In particular, we engaged experts at the Center for Sustainability at Saint Louis University. They asked us to personalize this core concept: What do people, planet, and prosperity/profit uniquely mean to Ameren? The answer: For people, our bottom line is to keep society running. For the planet, our bottom line is good stewardship. For prosperity, our bottom line is shared success going beyond profitability to include security, well-being, and community impacts.

"Saint Louis University's early involvement in our 2013 CSR report challenged us to think harder about what sustainability meant to Ameren, and to spend time articulating it in layman's terms," said Richard Smith, Director, Environmental Strategy, Ameren Corporation.[2]

---

[2] Personal Communication, February, 2017.

The structure of Ameren's CSR reporting shifted to mirror our approach to sustainability. We carefully weigh our responsibilities and priorities associated with key categories: sustainability at Ameren, customers and communities, workforce, the environment, and shareholders. The actions we take in one area can have both short- and long-term effects on others. Calibrating competing priorities across the categories is a constant challenge.

In 2013, it was challenging to create a report that more clearly described the key categories while also highlighting competing priorities and trade-offs of favoring one set of priorities over another. The CSR team was able to improve the overall approach to the report by making the following key improvements:

- Organizing our content according to the key categories
- Presenting the report in a digital format to conserve natural resources by publishing an interactive PDF and iBook for iPad available for download on our website
- Streamlining the content to about 50 pages, including full-page photos and interactive graphics
- Making it easier to find priority areas, facts, and figures and related resources online
- Introducing an interactive map of Ameren energy centers with relevant environmental data (e.g., location, capacity, generation, air quality and greenhouse gas emissions, water usage, coal combustion residuals and other key statistics)

The biggest optimization of 2013 involved candidly assessing Ameren's performance across the key categories—what we were doing well and what we could be doing better. Ameren's Corporate Sustainability Council helped in this effort. The Council, comprising top-level leadership and other representatives from across all areas of the company, was established in 2010 to help guide and grow our sustainability efforts. With a deeper understanding of our approach to sustainability, and why it matters to the quality of life, we developed a section called "How we are doing." It included an easy-to-read chart identifying areas of success and areas for improvement. The decision to include a self-assessment of our sustainability performance in the report reflected our strong desire to increase transparency with our stakeholders while also engaging our employees in more meaningful conversations about sustainability.

Our reporting period spanned January 1, 2012 to July 1, 2013; where appropriate, we included historical and/or forward-looking information for perspective. We collaborated with a communications firm, Simantel Group, Inc., to develop the report. It was issued in January 2014 (Fig. 2.3) and emailed to previously identified Ameren stakeholders, including media, legislators, and regulators. Before the report was distributed, we surveyed our internal audience to ensure the report would be valuable to their stakeholders.

Our second CSR report later won a Gold award from the League of American Communications Professionals (LACP) as one of the Top 40 Corporate Publishing materials in its 2014 Inspire Awards. There were more than 600 submissions from around the globe. The LACP is a national public relations industry group focused on providing a forum to share best-in-class practices.

Fig. 2.3 Ameren's second CSR report was published in 2014. Content was organized among five core areas of focus. This was also the first year with an interactive map of Ameren's energy centers. The popular feature provides specific, detailed information about each facility

2.3.1 Lessons Learned

Following this second report, Ameren employees were able to see the increased interest in sustainability reporting. Feedback from key stakeholders allowed internal groups to better understand the report's value and helped influence decisions. For example, Ameren employees collaborated on several important topics, including exchanging ideas on greening the supply chain, developing waste-recovery strategies and improving internal communications.

This report also confirmed the need for improved stakeholder outreach. Through previous conversations, we learned that an interactive map of our energy centers was an especially important element of the 2013 report. The data, which had always been available in a variety of regulatory filings, was presented in an easily accessible manner in the updated format. We found this helped both internal and external audiences learn more about our energy centers. The continued dialogue helped to further relationships with external stakeholder groups and aided in telling our story related to areas of key concern. Also, we received positive feedback from some stakeholders regarding the "How we are doing" section and realized this section should be included in future reports.

2.3.2 *Trade-off: Formal Reporting Standards Versus Company-Specific Approach*

A trade-off that we had to consider in this reporting cycle was whether we should strictly follow formal reporting standards or create a report that adapted frameworks to focus on topics relevant to our stakeholders. The GRI standards were the first global standards for sustainability reporting on a wide range of economic, environmental, and social topics. In our view, the GRI guidelines focused heavily on reporting using a template approach. During our CSR research, we noticed a lack of alternative frameworks, and we knew we wanted to focus on what matters to *our* stakeholders and what sustainability means to *our* company.

As with our first CSR report, Ameren followed the spirit of the GRI's latest guidelines and addressed main indicators contained in its Electric Utilities Sector Supplement. We felt confident in our decision to structure our report around the key categories, and crucially, we had internal support to dedicate the time and resources needed to develop our unique approach. We were able to dive into dedicated topics around each of our key categories. For example, for workforce, the topics were safety, culture, talent, and diversity. For shareholders, they were business model, leadership, value, risk, and ethics.

By adapting the formal reporting standards to fit Ameren's approach to sustainability, we were able to create a report that focused on topics relevant specifically to our stakeholders. The GRI standards provided a starting point for us to consider the important elements to include in a CSR report while excluding some of the information that would duplicate other disclosures already required by certain regulatory jurisdictions.

2.4 Our Third CSR Edition (2015): Increasing Frequency

Strategically, we decided in 2015 to coordinate release of the CSR report with the company's Annual Shareholders Meeting. This began when Warner Baxter became the CEO of Ameren Corporation in April 2014. With a clear commitment to corporate social responsibility, he asked the CSR team to complete our next report so it would be available the day of the annual meeting the following April.

To achieve this deadline, Ameren once again formed a CSR team. This time, our guiding principles expanded to include:

- *Substantive achievements*: Avoid creating a long list of everything we have done.
- *Substantive gaps*: Highlight key areas where we have performance gaps we want to close and identify what we are doing to close them.
- *Transparency*: Demonstrate a balanced approach to providing information on key areas of our performance, even when it is not always positive.

Since Ameren received recognition for our 2013 report, we decided to continue with the interactive PDF format. Our general reporting period spanned January 1, 2014 to December 31, 2014; this was the first time Ameren had reduced time between CSR reporting cycles and marked the beginning of an annual CSR report production cycle. This was also the first time we were reporting current data aligned with the release date, and we learned to be more efficient in obtaining, compiling, and verifying data. We picked up efficiencies by being more deliberate in our planning and by improving processes. For example, report authors had a better understanding of which of the 8600 Ameren employees had access to key information. Returning to them, or to their successor in a few instances, greatly reduced the amount of time needed to collect data. As it became evident the reports would be produced on an annual basis, more rigor was applied to sourcing, documenting, and filing source information. Not only was this identified as a CSR report best practice, it could help with retrospective analysis.

Being efficient paid off and we were able to meet the deadline challenge: The 2015 report was released on April 23, 2015 (Fig. 2.4), in conjunction with the annual meeting. Again, the report was digitally circulated among key stakeholders, including media, legislators, and regulators. Releasing the CSR report the day of the annual meeting allowed us to announce the report to our shareholders and emphasize our commitment to corporate social responsibility.

Fig. 2.4 Ameren's third CSR was published on the day of the 2015 Annual Meeting of Shareholders. Using lessons learned from previous reports, a single editor was assigned to each of the five sections

This edition addressed a range of topics important to diverse stakeholder groups, including Ameren's plan to transition to a cleaner, more diverse generation portfolio in a responsible fashion, the company's investments to modernize its energy infrastructure and create long-term value for customers and jobs for the local economy, and Ameren's charitable giving and employee volunteer efforts.

2.4.1 Lessons Learned

On a project as large as the CSR report, version control can be a difficult, so we empowered section editors to maintain version control and ensure consistency in reporting style and tone. Before information was assembled, the working team assigned an editor to handle each of the sections: sustainability at Ameren, customers and communities, workforce, environment, and shareholders. Editors were responsible for soliciting content, curating the information, updating data from previous reports and ensuring support documentation. The result of the new working team structure was a more efficient process. Leaders and those working on the team knew each section editor, which streamlined communications.

We also saw an increase in the number of international investors viewing our report and contacting us for information. These inquiries—along with various stakeholder discussions on environmental stewardship and customer value—helped form key insights going into our next edition. We recognized a need to communicate our CSR messages more effectively in alignment with our corporate strategy and annual reports. As an energy company, we understood that some global investors were comparing us with similar companies across the world and that investor interest in environmental, social, and governance (ESG) matters was increasing.

2.4.2 Trade-off: Resources Versus Frequency

Moving our CSR reporting cycle from every 2 years to annually was an important milestone in the report's development. Developing our CSR report is a resource-intensive process that requires input from many subject matter experts and leadership. A trade-off that we had to consider was the number of internal and external resources required for an annual reporting cycle. We needed to balance the time and expense of production against the frequency of publication. Some Fortune 500 companies have entire teams dedicated solely to documenting and reporting sustainability efforts. That is not the case at Ameren. Each team member has a significant number of other responsibilities.

The value of CSR reporting is inherently linked to the transparency afforded to stakeholders interested in corporate sustainability issues. We found that presenting sustainability information in an organized, company-specific manner is an excellent way to enhance communications with our shareholders, customers, and employees regarding Ameren's corporate responsibilities and priorities. CSR reports are also useful for creating positive dialogue and providing accurate information to other interested parties. Finally, we find that CSR reporting is extremely helpful for improving understanding of sustainability with employees across Ameren.

By weighing the resources needed to produce CSR reports against the benefits, we concluded that CSR reporting should be given priority and needs to be part of ongoing sustainability programs. It is difficult to quantify the monetary value of CSR reporting, but the communication benefits were clear. The decision to report annually was important to us, and we challenged ourselves to discover solutions to do more with limited resources—and we continue to evaluate process improvements and efficiencies. Additionally, the need for better, interactive, and timely CSR reporting led Ameren to move to a web-based platform.

2.5 Our Fourth CSR Report (2016): Digital Transition

We continued improving our CSR report by enhancing the user experience through creation of a digital report hosted on Ameren.com. This approach also reduced the resource needs for future reports. Data analysis of site visits to the report on Ameren.com, as well as focus groups and other market research, showed that users wanted to interact with Ameren on their device of choice, including smart phones. Investors, including those overseas, want readily available data. All audiences expect search engine results to generate the information they are seeking. This was best accomplished as a website and not as a document posted online.

Our new format—a microsite, AmerenCSR.com (Fig. 2.5)—also provided a fresh chance to rethink our content. We asked questions such as: Can we do more to illustrate the concept of balancing competing priorities among the key categories of sustainability? Can we improve our sustainability performance assessments? Where can we realize efficiencies by using existing assets, like video? Ameren's steering committee retained the core subject areas and restructured individual sections to further enhance clarity in each area of the report:

Why Ameren has a responsibility to this area.
What we are doing to show that.
How we are performing.
Who stands to benefit from our efforts.
See a featured topic of interest.

We gave the previous "Sustainability at Ameren" section in the previous PDF versions a new name: "Our Approach." Here, we leveraged lessons from our 2015 report: directly connect Ameren's approach to sustainability with our mission, *To Power the Quality of Life*; our vision, *Leading the Way to a Secure Energy Future*; and our business strategy, We will invest in regulated energy infrastructure, continuously improve performance, and advocate for responsible policies to deliver superior customer and shareholder value.

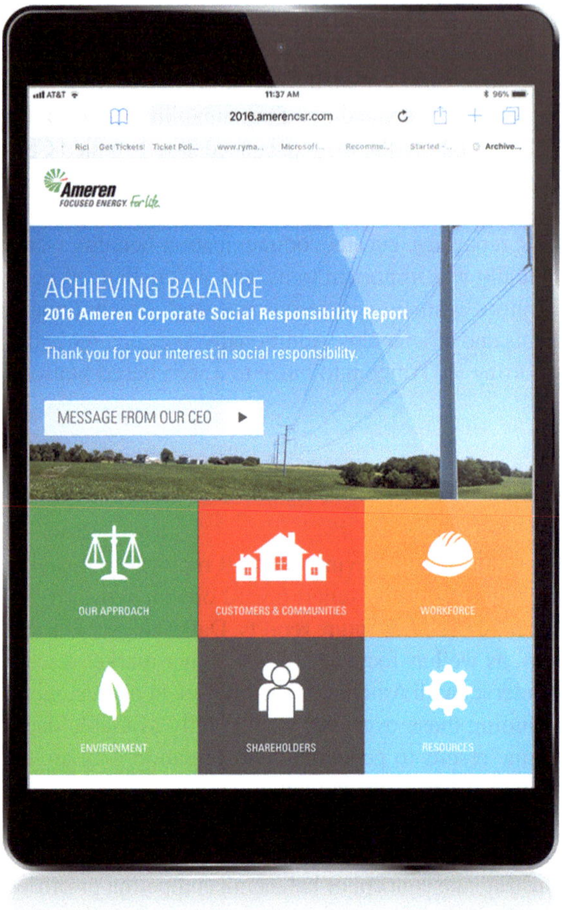

Fig. 2.5 The 2016 CSR Report was the first to employ a microsite that provided new insights into visitor behavior. This was the first report to be published exclusively online

Not only did we articulate how sustainability connects to these sections, we built an animation around the concept of "achieving balance." The animation was developed to walk users through our social responsibility in four key areas: customers and communities, workforce, environment, and shareholder. When making short- or long-term plans, we must weigh and carefully balance multiple priorities to help to make them truly sustainable. Certain actions in one area of responsibility can impact others. For example, Ameren is transitioning to cleaner energy, but this must be accomplished without compromising reliability. Renewables such as wind and solar are intermittent resources, so, it takes a diverse generation portfolio with dependable "baseload" capacity to meet our customers' needs 24/7. There are costs associated with planning, permitting, and building new-generation infrastructure. It is important to "achieve balance" as we transition to cleaner energy (Fig. 2.6).

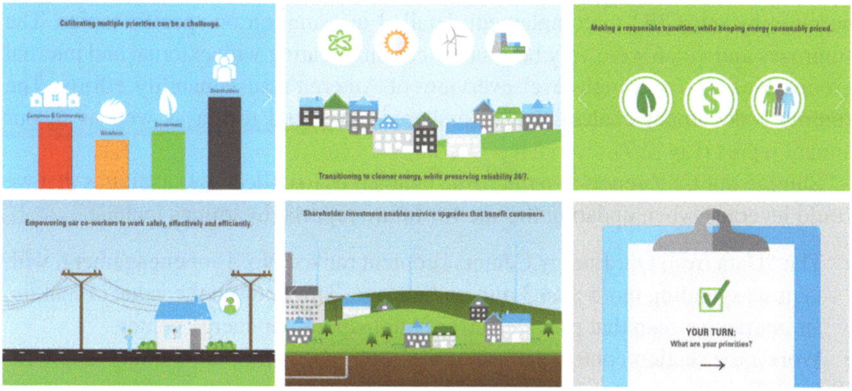

Fig. 2.6 Our 2016 CSR report, titled "Achieving Balance," brings this concept to life with an HTML animation

Clearer priorities were another notable update for 2016. In looking at GRI's latest available guidelines, the G4, we noticed the organization was encouraging companies to include reporting content aligned with the following [3]:

- Reflect an organization's "significant economic, environmental and social impacts."
- "Substantively influence" the assessments and decisions of stakeholders.

We consulted EPRI's report, *Material Sustainability Issues for the North American Electric Power Industry*, released April 2013 [4], to aid us in defining reporting priorities. After considering many possibilities, the CSR team ultimately articulated priorities for our customers and communities, workforce, environment, and shareholders that are similar to those contained in EPRI's technical report. Our 2016 CSR update also included summaries of how we performed against our priorities.

The CSR report microsite at AmerenCSR.com was launched on April 28, 2016, the day of the Annual Shareholders Meeting. Later that fall, our 2016 report was recognized as one of the Top 100 Reports Worldwide, ranking No. 20 out of nearly 1000 companies in LACP's Vision Awards. It was ranked No. 4 across all companies in the Americas Region and was the top-ranked report among US energy providers in the global competition.

2.5.1 *Lessons Learned*

Although we kept streamlining our reports with each cycle, we noticed that people wanted easier access to the information they cared about. For 2016, we saw an opportunity to provide salient content in alternate formats. We created a two-page downloadable executive summary with report highlights, as well as a brief video narrated by our CEO to complement detailed information on specific topics. The summary and video were very helpful for communicating with external and internal groups looking for a high-level overview of Ameren's sustainability efforts. The summary document content was also shared via social media as well as in our annual report (Fig. 2.7).

Staging our CSR report as a microsite enabled us to collect user analytics that we could leverage when updating the site for future reports. Learnings include:

- The "Data from Our Energy Centers" content ranked No. 1 for engagement, with visitors spending more than 3 min on the page. This showed the value of making an interactive map that provides information about our energy centers.
- Workforce section content ranked in the Top 5 pages for both number and duration of visits.
- Ameren employees accounted for 42% of initial site traffic. This signified that our effort to share CSR content with employees in internal channels was worthwhile.

Much of the microsite traffic for the 2016 report came from internal audiences. It was good for employees to hear about our sustainability actions and our approach to balancing the priorities of our key categories of sustainability. We believe

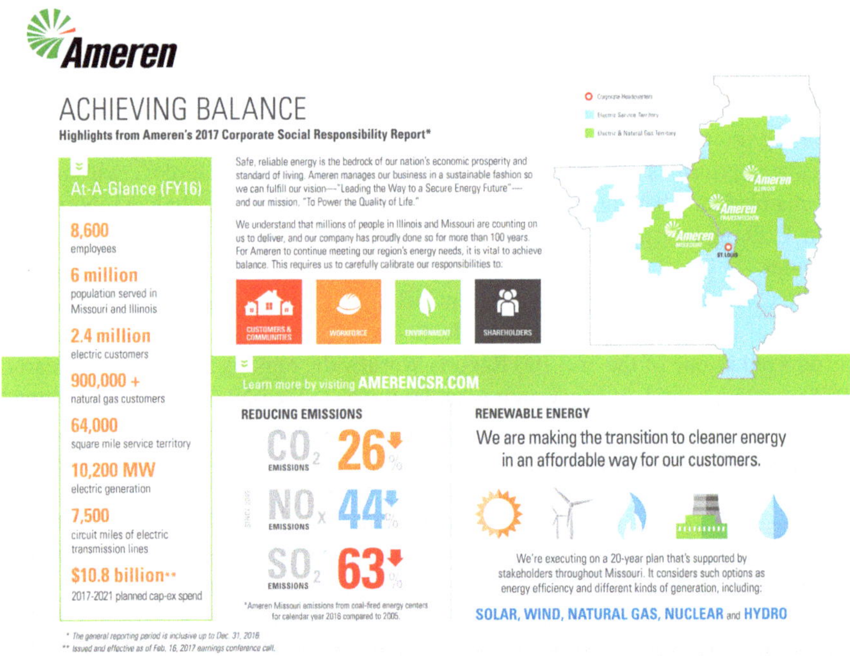

Fig. 2.7 The 2016 CSR Report summary is a quick look at some of the key statistics and information contained in the report. These images were also used in Ameren's Annual Report as well as the company's social media channels

employee engagement in sustainability efforts is key to successfully achieving our goals. Providing information through the CSR report is a good method to communicate information and insights across the company.

As always, we are looking for improvements. An identified improvement area is to reach out to more potential stakeholder groups so we can gain their perspective on environmental stewardship, customer value, and sustainability. Our strong desire to engage stakeholders on Ameren's unique approach to sustainability carried over into the real-world organization of, sponsorship of, and participation in the Achieving a Sustainable Energy Future Conference in April 2016 at Saint Louis University's Center for Global Citizenship. The conference brought together leaders from industry, academia, government, and the public to open the dialogue about sustainability and our energy future. It also provided evidence that we were addressing the key concerns of our stakeholders in our reporting efforts.

2.5.2 *Trade-off: Annual Website Update Versus Living Website*

Now that Ameren's primary CSR publication has an online home, we have cause to consider how the medium could shape the message in years ahead. A trade-off that we had to consider was whether to keep producing annual website updates versus maintaining a living website (i.e., updating our website year-round with even more timely CSR-related stories, videos, and news).

Additionally, technology enables more dynamic integration with Ameren's other online channels. At the same time, a good deal of CSR-related engagement is still happening through offline channels (bill inserts and mailings, speeches and presentations, and community events and conferences).

The CSR report is intended to be a snapshot in time of the company's actions related to sustainability and for the time being remains an annually updated report. This decision was based on resource availability and the desire to avoid the potential for confusion created by publishing new information alongside older information. The annual update philosophy also aligns with timing the release of the report with the Ameren annual meeting of shareholders. The decision was made to produce an annual CSR website update—congruent with the timing of key financial disclosures.

2.6 Our Latest CSR Website (2017): Refining Content

After devoting considerable time and resources to setting up the microsite and formatting of the 2016 version, the decision was made early in the project to update information on the website for the 2017 report rather than reinvent the format. The design team took on the challenge of giving the report a minor face-lift to improve navigation.

In planning the 2017 update, leadership expressed the need to include Ameren's focus on innovation. We are keenly aware that the expectations of our customers continue to rise at a time when energy technologies are advancing rapidly. There is no better time than now to focus on innovative investments. As a result, the company's efforts to embrace innovation are highlighted in each of the key areas.

Planning for the 2017 update also felt easier based on the team's experience with prior reports. Now that the report was being prepared on an annual basis, only a brief amount of time passed from the prior report's publication. Past decisions and choices the team made were still fresh in everyone's minds.

Also, a redesigned landing page took into account past lessons learned (Fig. 2.8). The front page highlighted areas that were important to us as a company, and referenced sections of the report we knew held key interest for our stakeholders.

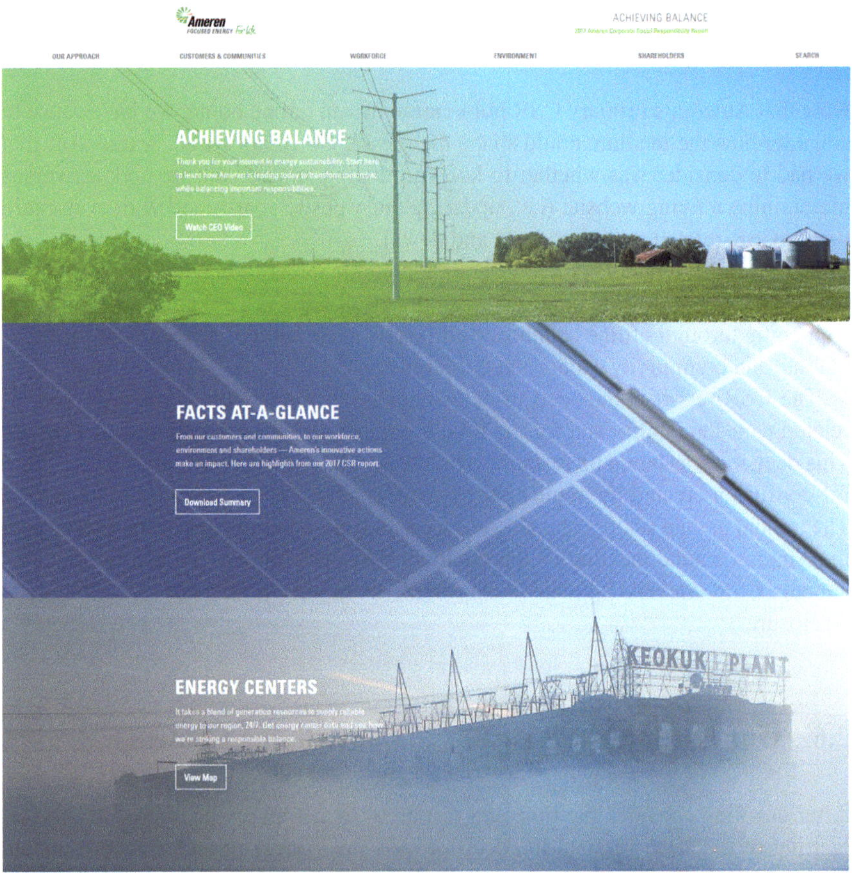

Fig. 2.8 The 2017 CSR Report refreshed the image of the established microsite and made a stronger link to the company's efforts on promoting and fostering innovation

2.6.1 Lessons Learned

Do not reinvent the wheel. This common expression drove a number of significant decisions in populating content in the 2017 CSR update. To be sure, there was a lot of new information and content to assemble, but each section editor first looked for Ameren stories that had already been told for another purpose. Existing video channels were combed for appropriate clips. Press releases were reviewed. These steps were valuable for two main reasons: They reduced the amount of time necessary to create content, plus the information was previously approved for use in a public forum. As a result of this work, section editors have created their own systems for flagging material for the next year's update. There is now a shared feeling that the ongoing review and capture of public information during the year will pay off with a significant time savings when it comes time to assemble information for future updates.

2.6.2 Trade-off: External ESG Verification Versus Internal Review and Recordkeeping

This year, much time and consideration was given to potentially using a third-party reviewer to verify data contained in our CSR report, in addition to our robust internal CSR recordkeeping practices. Proponents of third-party ESG verification claimed that external review adds credibility to CSR reports, which would be helpful to stakeholders interested in sustainability. We investigated potential approaches to ESG verification and learned that such efforts would greatly expand the schedule and cost of producing our report. At the same time, a survey of peer utilities revealed no consensus on whether using an independent external reviewer would provide enough value to justify the extra time and expense. Subsequently, the decision was made to continue with our disciplined recordkeeping process without engaging in external verification until such future time that the value of doing so becomes clear.

Achieving Balance in Practice: Real-World Tradeoffs
Ameren's CSR communications provide vital insights into our business decision-making process. The CSR provides a platform to share our approach to sustainability—achieving balance. As an investor-owned energy provider, Ameren has a social responsibility in four key areas: customers and communities, workforce, environment, and shareholder. When making short- or long-term plans, we must weigh and carefully balance multiple priorities to help to make them truly sustainable. While it is critical to talk about achieving balance among competing priorities, it is equally important to show how balance is pursued and achieved in practice.

Transitioning our Fuel Mix to a Cleaner and More Diverse Portfolio

Our customers want cleaner energy. Ameren is committed to transitioning our generation fleet in a responsible manner to a cleaner, renewable, and more diverse portfolio. We are executing on a 20-year energy plan that is supported by stakeholders in Missouri, where we operate a vertically integrated generation and energy distribution utility.

Our planning considers different kinds of electricity generation such as renewables, natural gas, and nuclear. It also considers factors that affect our decisions, including costs, risks, fuel diversity, customer preferences, economic development opportunities, and environmental performance. At the same time, we are embracing a culture of innovation, testing new technologies and materials that will help us achieve our goals.

As we continue this transition, our long-term plan is designed to provide service that is environmentally responsible, safe, reliable, and reasonably priced for customers. That is quite a balance to strike. For example, there are costs associated with planning, permitting, and building new-generation infrastructure, and additional transmission may be needed to carry renewable energy over long distances to urban and suburban population centers. Also, renewable resources such as wind and solar are intermittent; these are not guaranteed 24/7. A CSR report provides a mechanism to describe these competing issues and interests that must be balanced.

Given these and other considerations, Ameren is advocating for supportive energy policies that enable a balance of reliability, affordability, and sustainably sourced energy. Meanwhile, we have significantly driven down emissions at our existing energy centers. To further aid our transition while balancing trade-offs, Ameren is:

- *Expanding customer energy efficiency programs*. These currently represent the lowest-cost method for our customers to manage their energy use, and we embrace them. Since the programs were introduced in 2008, Ameren Illinois customers have saved about as much electricity as the annual consumption of about 966,000 average-sized homes; since the introduction of new programs in 2009, Ameren Missouri customers have saved the equivalent of taking 156,000 homes off the grid a year. Ameren is planning to spend over $335 million on energy efficiency programs in Illinois and Missouri through 2018.
- *Expanding cleaner and diverse energy resources* by building additional renewable energy generating facilities to meet customer demands as fossil-fuel facilities reach the end of their useful lives (Fig. 2.9). In addition, Ameren Illinois and Ameren Missouri have short- and long-term renewable power purchase agreements, supporting renewable energy in the Midwest.

Fig. 2.9 Ameren Missouri's O'Fallon Renewable Energy Center. Put in service in 2014, the 19,000 multi-crystalline solar panels span an area approximately the size of 19 football fields. It allows us to collect nearly 8 million kilowatt-hours of clean energy a year

- *Retiring coal units as they reach their normal end of life* to minimize customer cost impacts. For decades, coal-fired plants have produced reliable energy around the clock. As these plants age, they cannot be easily replaced. Our company is currently planning to integrate newer, cleaner energy sources, as the technology matures and becomes more affordable, into our smarter energy grid in the future to benefit our customers. Coal generation will continue to represent a smaller percentage of our rate base, accounting for less than 12% of estimated rate base in 2022.

"As we transition our generation portfolio, we look to innovate and achieve emission reduction targets primarily through flexible and innovative operational measures—instead of through cost-intensive, traditional environmental controls," said Ajay Arora, Vice President of Power Operations and Energy Management, Ameren Missouri. "Ameren is confident that we can achieve both environmental and financial stewardship, managing costs for the benefit of our customers and shareholders."[3]

[3] Personal Communication, March, 2017.

2.7 Conclusion

As our CSR reporting journey has evolved, we have learned a few key insights: Deliberately clarify your company's approach to sustainability, engage external and internal audiences, make relevant information easy to find, and present it in an understandable way.

The report continues to evolve. The journey started with a desire to find a better way to communicate what was very important to us: our efforts surrounding social responsibility. A major aspect of that message is our sustainability efforts surrounding our goal of achieving balance in key areas: our approach, customers and communities, workforce, environment, and shareholders.

There have been many opportunities to assess which direction we wanted to take the report. The largest of the many trade-offs we encountered evolved as follows:

- *More Reporting vs. Not.* The decision was not easy, but ultimately to start this process, it was necessary to embrace the concept that information previously disclosed to the public in various filings or disclosures would be collected and crafted into an easily accessible format for all stakeholders.
- *Formal Reporting Standards vs. Company-Specific Approach.* For Ameren, existing CSR reporting standards offered a starting point for us to build our own custom approach. This allowed us to highlight and focus on what matters to *our* stakeholders and what sustainability means to *our* company.
- *Resources vs. Frequency.* This is a resource-intensive process, but the promise of significantly better dialogue with stakeholders led to the decision to produce an annual CSR report, which helps communicate information in a transparent and timely manner.
- *Annual Website Update vs. Living Website.* Technology allows for constant updates, as does the online-only format. Resource allocation helped guide the decision to report once a year. Reporting at the same time as the annual meeting allows us to provide more information to stakeholders when they are most likely to be interested.
- *External ESG Verification vs. Internal Review and Recordkeeping.* The decision was made to continue with our disciplined internal information review and recordkeeping process without engaging in external verification until such future time that the value of doing so becomes clear.

Although it is difficult to quantify the monetary value of CSR reporting, we believe that our voluntary reporting efforts have:

- *Engaged key stakeholders on issues and the need to balance multiple priorities across many business activities*, allowing us to continue the conversation on key energy matters.
- *Helped advance internal thinking* and focus on sustainability and our approach to balancing sustainability priorities.
- *Provided accurate, publicly available information* and resources to interested parties, investors, and investor advisory service firms in a sustainability context.

- *Received external recognition/awards* among our peers and other industries; this pushes us to keep raising the bar on our reporting format.

As we look to the future, our customers continue to expect that energy companies like Ameren will deliver safe, reliable, and cleaner energy at a reasonable price. It can be challenging to balance the needs of all our stakeholders, yet we are always focused on the fact that our customers count on us, 24/7. Meeting and exceeding their expectations will continue to be key to our overall success. Ameren is committed to carefully balancing all our responsibilities in a sustainable way as we work to fulfill our mission to *Power the Quality of Life* today and our vision of *Leading the Way to a Secure Energy Future* for generations to come.

Author Profiles

Ajay Arora Ajay Arora is the Vice President of Power Operations and Energy Management. Arora is responsible for managing operations for Ameren Missouri's diverse energy generation portfolio, energy management and trading, environmental policy advocacy and compliance, as well as operational excellence. Prior to this position, Arora was the Vice President of Environmental Services and Generation Resource Planning. He has also directed Corporate Development, Market Risk Management, and briefly the Corporate Project Oversight groups at Ameren. Arora graduated from Tulane University in New Orleans with a Master's degree in Business Administration and from Punjab University with a Bachelor's degree in Chemical Engineering.

Brad Brown Brad Brown is the Environmental Communications Manager at Ameren Corporation. He is responsible for developing and implementing internal and external communications strategies around federal and state environmental regulations related to Ameren's business, including Ameren Missouri's energy generation portfolio objectives. He also leads communications related to Ameren's growing renewable energy initiatives. Recent projects include Ameren's Corporate Social Responsibility Report and Ameren Missouri's Integrated Resource Plan. Brown earned a Master of Arts degree with honors from Webster University. He also holds a Bachelor of Journalism degree from the University of Missouri.

Diane Saftic Diane Saftic is a supervising engineer with Environmental Strategy at Ameren Corporation. Saftic focuses her responsibilities on environmental and sustainability issues. Saftic earned a Bachelor of Science degree in chemistry from the University of Kansas and a Master of Science degree in environmental engineering from Washington University, St. Louis. She is a registered professional engineer in Missouri.

Richard Smith Richard Smith is the Director of Environmental Strategy at Ameren Corporation. He has over 40 years of industrial experience including engineering, manufacturing, development and management of major projects, electricity generation operations, business planning, advanced technology management, environmental services, and sustainability leadership. Smith is a fellow of the American Society of Mechanical Engineers and is a certified Project Management Professional.

Smith is licensed as a Professional Engineer in the States of Missouri and Illinois. He earned Bachelor and Master of Science degrees in mechanical engineering from the University of Missouri, Rolla.

Steven Whitworth Steven C. Whitworth has over 37 years of experience in the electric utility industry. Whitworth has worked in various capacities in power generation, including engineering, operations and maintenance, as well as managing environmental issues at both the facility and corporate levels. Whitworth has provided thought leadership for the development of Ameren's sustainability initiative. In his current capacity as Senior Director, Environmental Policy and Analysis, Whitworth directs development of environmental policy and leads advocacy efforts on environmental issues. Whitworth has overall responsibility for Ameren Missouri's environmental services function, including coordination of compliance strategy development and environmental support for Ameren Missouri energy delivery and generation. Whitworth has degrees in Biological Sciences and Mechanical Engineering Technology.

Company Profile

St. Louis-based Ameren Corporation powers the quality of life for 2.4 million electric customers and more than 900,000 natural gas customers in a 64,000-square-mile area through its Ameren Missouri and Ameren Illinois rate-regulated utility subsidiaries. Ameren Illinois provides electric distribution and transmission service, as well as natural gas distribution service, while Ameren Missouri provides vertically integrated electric service, with generating capacity of nearly 10,300 megawatts, and natural gas distribution service. Ameren Transmission Company of Illinois develops regional electric transmission projects. For more information, visit Ameren.com.

References

1. 2011 Ameren Corporate Social Responsibility Report.
2. Global Reporting Initiative. "About Sustainability Reporting." *Globalreporting.org*. www.globalreporting.org/information/sustainability-reporting/Pages/default.aspx. Accessed 06 March 2017.
3. Global Reporting Initiative. "Materiality." *G4.globalreporting.org*. g4.globalreporting.org/how-you-should-report/reporting-principles/principles-for-defining-report-content/materiality/Pages/default.aspx. Accessed 06 March 2017.
4. *Material Sustainability Issues for the North American Electric Power Industry: Results of Research with Electric Power Companies and Stakeholders in the United States and Canada*. EPRI, Palo Alto, CA: 2013. 3002000920.

Chapter 3
Reclaiming Value from Former AEP Mine Lands: Balancing Economic and Environmental Benefits

Timothy W. Lohner, C. Michael Williams, and Jeffrey W. Wilson

Abstract When conflicts arise, sometimes compromises must be made to balance resource extraction from reclaimed mine lands for economic benefit while reestablishing ecological communities and functional ecosystems. The resulting narrative encompasses mine land reclamation, forestry practices, natural gas extraction, and ecosystem services, such as species conservation, carbon sequestration, and water management. The choices that are made are tradeoffs – giving up one thing to get more of another. In the effort to protect the environment and natural resources, the challenge comes in finding approaches to environmental management that give people the quality of life they seek while protecting environmental systems that are also foundational to society's well-being.

3.1 Introduction

Imagine you discover a magical woodland that contains exotic wildlife, colorful fish, and quick-growing bushes. There are strange black rocks that burn and a mysterious gas from deep underground that glows with a blue flame. Something like the lands of Narnia [1] or Terabithia! [2] You and others explore and enjoy the fields and forests. You go fishing, hiking, camping, and boating. Then one day you learn that you have to make money from this property. Now you have a problem. Do you cut down the trees, build roads, and drill to extract and sell the gas? Do you sell the timber and keep the property? Or do you sell everything? What about the fish, deer, and other wildlife? How do you balance these options and accommodate everyone? Tradeoffs will be made, and not everyone will like them.

While this may sound like an imaginary story, this is exactly the situation at the AEP ReCreation Land in southeastern Ohio. This 60,000-acre property was originally acquired for mining readily available coal deposits for consumption at the company's nearby Muskingum River Power Plant. After the mining, the area was

T. W. Lohner (✉) · C. M. Williams · J. W. Wilson
American Electric Power, Columbus, OH, USA
e-mail: twlohner@aep.com

J. Fox, M. Scott (eds.), *Sustainable Electricity II*,
https://doi.org/10.1007/978-3-319-95696-1_3

“reclaimed” by AEP, which planted more than 63 million trees, nurtured thousands of acres back to productive pasture, and established 350 lakes and ponds, stocking many of them with fish. It also established 380 campsites and created hiking and horseback riding trails. A 10,000-acre tract of AEP ground adjacent to ReCreation Land was even donated to the Columbus Zoo and Aquarium to create The Wilds, a private, nonprofit safari park and conservation center where rare and endangered species from around the world now live.

But today, there is competition for how the site should be used. Some want continued public access for hunting, fishing, hiking, and camping opportunities. Still others, including AEP, are pressured to obtain economic value from the property. Many acres are leased to farmers, who use the land to pasture livestock or grow crops. Timber is sustainably harvested, and tracts of ground on the fringes of the ReCreation Land site are being sold for vacation homes and recreational purposes. While oil and gas exploration was common even before AEP acquired the land, more recently leases have been negotiated to allow the extraction of oil and natural gas using the new “hydraulic fracturing” technique, which is often called “fracking.” Looking into the future, large portions of the property may be used to create solar farms.

Land use decisions today involve choices: choices between environmental and economic uses. Compromises are often necessary, yet people can become angry when they don’t get what they want. While many environmental benefits – such as wildlife viewing, hunting, fishing, biodiversity preservation, and watershed management – may be realized, they come at a cost. The property that makes up AEP’s ReCreation Land was originally purchased to provide access to coal, which was used to generate electricity, which in turn was sold for a negotiated profit. While there are obvious public relations benefits to maintaining the outdoor recreational opportunities at ReCreation Land, AEP is not in the business of running parks and wildlife sanctuaries. Instead, its primary responsibility is to generate a safe and reliable source of affordable electricity.

Conflicts have arisen between those supporting ecological protection and those wishing to extract or economically utilize on-site natural resources. As a result, negative press has been generated [3, 4].

This chapter will describe how tradeoffs have been made to accommodate the goals of all stakeholders, including those with economic interests as well as those wishing to conserve the ecological value of the property. There is much to consider on this topic, and lessons have been learned, but it can be difficult to maintain this balance. Still, it may not be too late to keep that magical wonderland that all can enjoy and be proud of.

3.2 Site Description, Location, and History

AEP ReCreation Land is located south of Interstate 70 in a rural area near McConnelsville, Ohio. The approximately 60,000-acre property includes portions of Muskingum, Morgan, Noble, and Guernsey counties and exists within an

84,000-acre ownership stretching over five counties known as the Muskingum Mine (Fig. 3.1). This land was historically used for agriculture until the 1940s. Morgan County is an economically disadvantaged part of the state where more than 21% of the population lives below the national poverty level [5] and the median household income in 2015 was $37,067. The local economy is dependent on agriculture, forestry, coal mining, and, more recently, oil and natural gas extraction. An AEP subsidiary, Central Ohio Coal Company, began mining the property in 1947 and extracted more than 110 million tons of coal. The locally famous "Big Muskie," a huge dragline shovel that could move 220 cubic yards of earth in a single bucket, played a big part in this earth-scaping (Figs. 3.2 and 3.3). More than two billion cubic yards of earth – more than eight times the amount moved in the building of the Panama Canal – were moved to gain access to the coal.

Reclamation of the property was an ongoing process that began immediately following coal extraction and continued after the mining was completed in 2001. The land was backfilled and graded to direct water runoff to lakes and ponds. In the earlier stages, trees were planted, but later, following the Surface Mining Act of 1977, quick-growing grasses were seeded and established in compliance with coal mining reclamation laws. Typically, once an area has been mined, soils are left compacted, severely limiting the ability of tree seedlings to grow and survive. Since a mine operator must successfully establish a vegetative cover prior to the release of its reclamation bonds, the establishment of grasses is initially favored. The restora-

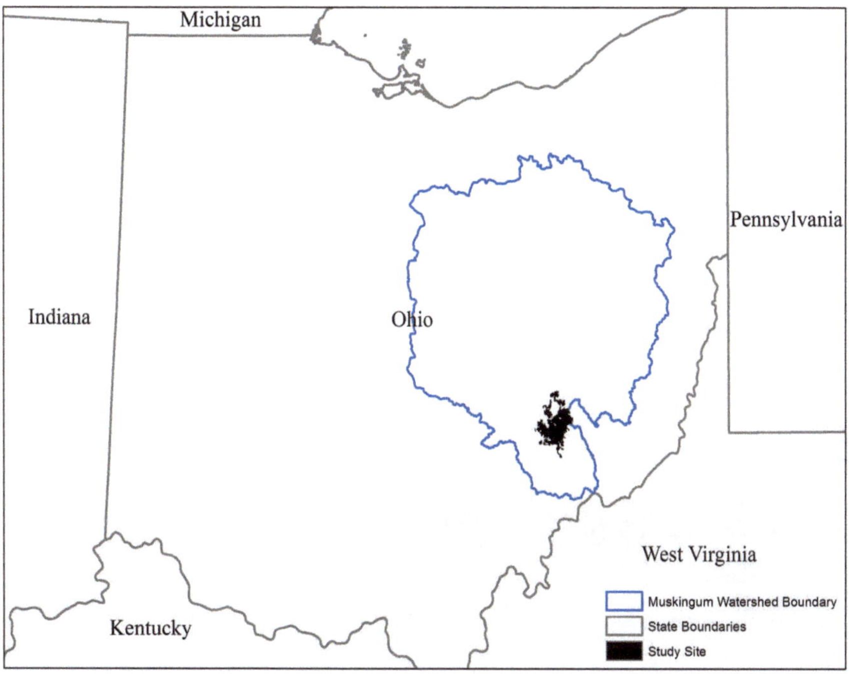

Fig. 3.1 Location of AEP ReCreation Land in Ohio

Fig. 3.2 The Big Muskie Bucket could hold two Greyhound buses side by side

Fig. 3.3 The Big Muskie dragline was powered by 13,300 volts of electricity

tion efforts at ReCreation Land have provided both recreational and ecological value as the site has matured over the past six decades.

3.3 Recreational Value

In many ways, the property does resemble a place where a kid, or even an adult, can have an adventure. There are opportunities to hike, fish, hunt, trap, canoe, horseback ride, mountain bike, or view wildlife, and access is free to the public. For many years, AEP has hosted an annual Wheelin' Sportsmen event at ReCreation Land, pairing disabled hunters with able-bodied volunteers in a controlled deer hunt. Dozens of local community members and AEP employees volunteer every year to provide opportunities for the disabled hunters to enjoy both group camaraderie and hunting experiences they otherwise might not have. AEP partners with many groups to support this activity, including the National Wild Turkey Federation, the Ohio Department of Natural Resources, and the Ohio Army National Guard. Since the event's inception, hundreds of hunters have been able to participate.

ReCreation Land is also home to many other educational/recreational programs, including Hooked On Fishing – Not On Drugs, a Career Day event for local schoolchildren, and pond clinics where local residents learn how to identify and control vegetation, develop fish stocking programs, and manage pond systems. AEP works with countless organizations to assist with these opportunities, including the Ohio Horseman's Council, the Columbus Astronomical Society, and the Ohio Department of Agriculture. More than 100,000 people visit ReCreation Land annually to enjoy these programming events and other recreational activities (Table 3.1, Figs. 3.4 and 3.5).

3.4 Ecological Value

AEP ReCreation Land was created during the mid-1960s, after restoration, primarily for recreational use by the public, hence the name. Geographically, the site consists of various habitat types, including floodplains, coves or ravines, sideslopes and ridgetops, spoil piles, relic highwalls, and reclaimed mine lands [6]. A large portion of the floodplain is used for agricultural production, while smaller areas are forested. The coves and ravines contain streams that flow throughout the site. Spoil banks and highwalls are features left over from surface mining that occurred before regulations required that they be removed and contoured. Spoil banks, or piles of coal mining debris, have been stabilized and support forest vegetation. The reclaimed portions of the site have been recontoured according to regulations and stabilized with grass and tree species, including aggressive vegetation such as tall fescue and lespedeza. Tree species include quick-growing sycamore, black locust, European alder, green ash, sweet gum, bald cypress, white pine, and pitch pine [6]. Species

Table 3.1 AEP ReCreation Land facts and figures

Acres	60,000+
Visitors	100,000+ annually
Organizations hosted	School groups, fishing clubs, Scout groups, Buckeye Trail Association, National Wild Turkey Federation events
Ponds and lakes	640
Hiking trails	28 miles (includes 24-mile portion of Buckeye Trail)
Horseback riding trails	30 miles
Mountain biking trails	10 miles
Campsites	380
Trees planted since 1944	63 million on AEP property, many at ReCreation Land
Selected awards	Governors Tree Source and Edmund Secrest tree planting awards Distinguished Service Award for the conservation of Ohio wildlife Ohio Tree Farmer of the Year Award Wildlife Habitat Council Certification Bill Miller and Land Owner Awards from Buckeye Trail Association National Land Management Award Blue and Gold Award from Muskingum County for community support Economic Activism Award from Morgan County Chamber Cardinal Award from the Ohio Department of Natural Resources Certificate of Appreciation for support of Central Ohio Mountain Biking Organization National Wild Turkey Federation awards for Wheelin Sportsmen hunting events

beneficial to wildlife were also planted, including oaks, hickories, and walnut. Timber harvesting has been done to promote the growth of these tree species, creating structural diversity in the harvested areas and increasing the diversity of habitat. During AEP's Climate Challenge Tree Planting Project, 15 million trees were planted, a large portion of which were located on ReCreation Land.

While many non-native tree species have been planted on the site, some species native to Ohio, such as red and white oak, black cherry, yellow poplar, flowering dogwood, and sugar maple, can also be found on the property [6]. These tree stands represent high-quality forests that can provide habitat for species such as the Indiana bat, American burying beetle, Kirkland's snake, and several species of songbirds, all of which are of special interest to biologists [7].

Stands of native trees can support biodiversity, which is the variety of life in a particular habitat or ecosystem. Forests improve overall habitat, conserve water resources, and sequester carbon dioxide, a greenhouse gas. In fact, many of the tree plantings by AEP were established to create tree "plantations" with the specific intent of sequestering carbon dioxide. At the time, it was believed that planting trees was one of the most effective and lowest-cost ways to remove the gas from the atmosphere [6]. However, we now know that it is difficult to guarantee that the carbon dioxide sequestered in the trees will remain permanently out of the atmosphere,

Fig. 3.4 Fishing event at ReCreation Land

Fig. 3.5 Lake near Sand Hollow Campground, ReCreation Land

because trees on a different property may be harvested or burned for fuel, eliminating the gains of the planted trees [8].

The benefits to humans provided by forests are also known as "ecosystem services." These are resources and processes that are provided by nature at no cost to us and that we would not be able to function without. For example, without pollination, there would be a complete failure of many agricultural crops. More than one-third of the world's crop species, including numerous nuts, fruits, and vegetables, depend on bee pollination, an ecological service valued in North America at $20 billion a year [9–11]. Insect-eating bats save the US agricultural industry at least $3 billion a year, yet bats are among the most overlooked, economically important, non-domesticated animals in North America [12]. ReCreation Land supports such species and many other vital ecosystem services.

In contrast, some species at ReCreation Land are not particularly beneficial because they can take over and be harmful to their surrounding habitat. The non-native, invasive, and aggressive autumn olive and multiflora rose were planted on the site in order to achieve quick reclamation of the property. Autumn olive, in particular, was planted due to its ability to dominate open areas and create a quick habitat and food source for wildlife. However, when any activity or tree harvesting opens up an area, this species will quickly establish itself, preventing more valuable species from regenerating.

The desire to reclaim these properties and establish viable ecological communities is illustrated by recent work done by the Ohio Department of Natural Resources (ODNR). This agency purchased four large tracts of strip-mined land very near the AEP property with the intent of establishing native grassland bird species (Fig. 3.6). The former mining areas represent thousands of acres of unbroken grassland habitat, something that is not generally available in Ohio. However, reestablishing grassland bird species presents many challenges, mostly related to the presence of invasive species, which choke out more desirable plants. The ODNR devoted many resources to this challenge, including applying special herbicides by helicopter and mulching the dead woody vegetation. The ODNR has featured this work in its seasonal publication, *Wild Ohio*, in articles such as "Reshaping the Land for Wildlife" [13] and "Coal Mine to Gold Mine" [14].

RESHAPING THE LAND FOR WILDLIFE

Ryan Harris, Assistant Wildlife Management Supervisor for the Ohio Department of Natural Resource's (ODNR) Division of Wildlife, noted that when the agency purchased large tracts of formerly strip-mined lands in southeastern Ohio, it was presented with a unique opportunity to establish new grasslands. Such habitat is not common in the state, and certain bird species can benefit from this environment. The ODNR targeted six bird species – bobolink, eastern meadowlark, grasshopper sparrow, Henslow's sparrow, prairie warbler and northern bobwhite quail – in an effort to restore the nearly 11,000 acres of grasslands. This involved the removal of invasive woody species, such as autumn olive and black locust, through aerial herbicide application and the use of specialized mulching equipment, followed by the planting of native grass seeds. This work has been successful, as demonstrated when a group from the Central Ohio Chapter of the Ohio Young Birders Club observed Henslow's and grasshopper sparrows during 2017 at the Tri-Valley Wildlife Area, where 150 acres had been improved two years prior. (Harris, Ryan. "Reshaping the Land for Wildlife." Wild Ohio, Jan./Feb. 2015. pp. 20-21.)

Fig. 3.6 The Ohio Department of Natural Resources is re-establishing grassland bird species

3.5 Conflicts and Tradeoffs

The economic activities conducted at ReCreation Land can impact the environment in ways that may be at odds with those who prefer to see the property preserved for its ecological and recreational benefits (Table 3.2). Oil and gas development, timber harvesting, pasturing, crop production, real estate sales, and future solar development can all have a negative impact on biodiversity and ecosystem services, rare or endangered species, water resources, and the recreational value of the property. On any path, there are tradeoffs that have economic as well as ecological consequences.

The issue of environmental tradeoffs is complex and often includes issues that are not always obvious, particularly when they involve extractive industries and biodiversity. While the ecological and biodiversity losses are often dramatic and receive a great deal of attention, what is not mentioned is how the economic benefit provided by the extractive activity benefits local communities. Balancing the needs of the local economy with the preferences of those who favor ecological protection is an ongoing challenge for AEP, one that the company, up to this point, has successfully negotiated. AEP works with the local communities to find the best compromise for the use of the property, whether it be sustainable timber harvesting, pasturing, or limited property sales. These activities, in turn, benefit not only AEP and its shareholders but also the local residents through jobs, taxes, and the support of local businesses. But if a community's basic needs are not met, it will not see ecological protection as a worthwhile goal. Instead, the environment may be seen as a resource to exploit. Illegal dumping and poaching are often problems in economically depressed rural areas. The pages that follow will describe the economic, environmental, and social pressures associated with each property option and how they have been and are currently being addressed by AEP.

Table 3.2 Environmental and social threats due to ReCreation Land development

Threats	Distribution and intensity	Legal/social/ physical protections	Habitat sensitivity
Oil and gas development	Wells throughout site being drilled, approximately 407 acres; cleared pipelines	Oil and gas regulations, lease agreements	Water (salinity, total dissolved solids, chlorides), endangered species, loss of wildlife habitat, air and noise pollution
Timber harvesting	Throughout site, approximately 300 acres annually	State regulations, lease agreements	Water siltation, endangered species, loss of wildlife habitat
Pasture	5222 acres	Lease agreements	Water siltation, waste and nutrient runoff
Agriculture	2862 acres	Lease agreements	Water siltation, waste and nutrient runoff
Solar development	1000 s of acres	State/federal regulations	Loss of wildlife habitat and recreation (hunting, hiking, etc.), increased runoff, bird strike impacts

3.5.1 Sustainable Forestry

After first acquiring the property in anticipation of mining, AEP harvested all merchantable timber ahead of mining operations, marketing this asset to timber companies and paper mills. Timber harvesting can result in the loss of forest functions, such as water infiltration, stability of local climate patterns, soil fertility, carbon sequestration, and biodiversity [15]. Forests also provide cultural services, such as recreational and "spiritual" experiences. Think about the wonder that is the gigantic size of the redwoods in California or the sight of nesting bald eagles to gain perspective on the impact nature has on culture. The loss of such trees or wildlife would be devastating to those who value them for more than their economic value. Unsustainable land management, in particular, threatens biodiversity, forest resources, and ecosystem services.

In an article in the scientific journal *Forest Policy and Economics* [16], it is noted that the tradeoffs surrounding forest management are often centered on two human-altered systems rather than between a man-made system and a natural one. In other words, the forest in question is not a virgin forest but one that has been managed for timber, and the alternate uses may be cattle grazing or agriculture. The issue of forest management at ReCreation Land is no different. The ReCreation Land property has been altered from its original forested condition for more than 100 years. Well before AEP mined coal at the site, the property was altered to grow crops and pasture cattle. Cattle grazing, agriculture, and plantation forestry have different effects on water quality and quantity, soil conditions, carbon sequestration, biodiversity, and other conditions [16]. While some stakeholders may not want the trees to be harvested at all, doing so provides a value by reducing the pressure on neighboring forests. By managing its forests for tree harvesting, AEP is indirectly discouraging foresters from harvesting trees from other forests, which may have unique ecological value.

Today, AEP manages the vast wealth of timber on the site, conducting annual, sustainable harvests and carrying out timber stand improvement activities (Fig. 3.7). In order to preserve the functional value of the forests, the company created a Forest Management Plan, which prescribes sustainable management of the timber reserves on the property. For example, the timber that was planted as part of the reclamation effort was thinned once, and in some cases twice, to encourage the healthy growth of the trees [17]. This sustained management of the forestlands has preserved the ecological and recreational value of the property, avoided potential conflicts with stakeholders, created jobs for the local economy, maintained forest and habitat diversity, established habitat for wildlife, generated revenue, and improved access to the property.

Conventional forest management considers only the volume of timber extracted. It does not consider uses, such as using the timber for biomass or carbon storage or protection of the local watershed. However, sustainable forest management like AEP's can generate proven synergies and ecological benefits [15]. Forest plantations, especially those with high growth rates, can add significant environmental benefits through "carbon sinks" [18]. The trees absorb carbon dioxide as they grow

Fig. 3.7 Sustainable timber harvesting at ReCreation Land

and respire, removing it from the atmosphere, serving as a "sink" or "reservoir" for the greenhouse gas. Another benefit is the harvesting of dead trees for firewood or other uses, which can reduce the intensity of any forest fires [15].

These forests have also been certified under the American Tree Farm System [19]. This program is internationally recognized and verified through the application of strict audit standards. The program standards outline best management practices, which address four key areas: water, wildlife, wood, and recreation. Examples of these standards include the maintenance of a forestry management plan, which outlines the owner's objectives to address forest health, soil, water, wood and fiber production, threatened and endangered species, invasive species, pest management, and high conservation values [19]. Property owners are also encouraged to consider opportunities to protect and enhance wetlands, as well as desired fish, wildlife, and plant species, recreation, aesthetic quality, and biomass and carbon [19]. As part of its commitment to environmental stewardship, AEP has been inspected multiple times by PricewaterhouseCoopers for compliance with the program and was recognized as Ohio Tree Farmer of the Year in 2000.

3.5.2 Managed Pasture Leasing

Several thousand acres of ReCreation Land are leased to local farmers who graze cattle and mow hay on the property (Fig. 3.8). These land use practices of pasturing and growing hay or alfalfa have their own potential impacts, possibly harming

Fig. 3.8 Hay mowing on leased property

nearby streams due to excess nutrients and sedimentation. Many farmers apply fertilizers, which include phosphorus and nitrogen, to their fields to increase crop production. If the fertilizers or the nutrients from livestock manure exceed the crop needs or have not been incorporated into the soil before it rains, they can drain into nearby streams and can cause harm [20]. When the nutrients are present in high levels, they can cause noxious algal blooms that impair aquatic life and threaten drinking water systems. The nutrient pollution can also lead to fish kills, oxygen-starved hypoxic or "dead" zones, and public health concerns related to increased exposure to toxic microbes such as cyanobacteria [21, 22].

The pasturing of livestock can also have direct impacts on the land and nearby streams. Cattle grazing near rivers can cause stream bank collapse, which is a major source of sediment. This sediment, in turn, can harm the life in the streams, smothering aquatic insects and preventing fish from spawning [23–25].

Rather than simply signing leases with farmers for pastures in exchange for financial gain, AEP, in cooperation with the local county Soil and Water Conservation Department, works with the farmers to establish comprehensive "grazing management plans." This arrangement benefits both parties and reduces the tradeoffs between the environment (water quality) and the community's needs (farming/livestock). These plans include requirements to implement management plans to ensure that cattle populations are appropriate for the land area and that steps are taken to protect ponds and watersheds and implement rotational grazing practices to prevent overgrazing, making the pasture leases beneficial to the newly reclaimed mining ground as well as to the local community and economy.

3.5.3 Natural Gas Extraction

Oil and gas exploration and development of the property have been a reality since before AEP's ownership. The first commercial oil well in Ohio was drilled in 1860 near Macksburg, in Washington County, not far from the AEP ReCreation Land site. These early wells extracted oil from shallow sandstone reservoirs in southeastern Ohio, but today most wells are drilled and developed through a technique known as "hydraulic fracturing" (Fig. 3.9) [26]. There has been a rapid increase in natural gas production from this drilling technique, in which a well is drilled down to the depth of the target reservoir and turned approximately 90 degrees to run laterally through the rock formation, usually shale rock. Due to the low permeability of shale, producers hydraulically "fracture" the rock by injecting water into it at tremendously high pressures. This creates cracks or "fractures" in the shale, through which the natural gas flows. To enable better flow of the gas, "proppants," or small sand-like particles, are added to the injected water, which keep the fractures open [27]. However, the environmental consequences of this process are being debated as the impacts of these new technologies are examined. Potential impacts include the contamination of groundwater, methane pollution and its impact on climate change, air pollution, exposure to toxic chemicals, water use in water-deficient regions, fracking-induced earthquakes, and waste disposal [27–29]. Although an estimated 80,000 wells have been fractured in Ohio, state agencies have not identified a single instance where groundwater has been contaminated by hydraulic fracturing operations [26].

AEP leased mineral rights for hydraulic fracturing to Anadarko Petroleum, a Texas-based drilling company, shortly after 2010. Anadarko originally proposed to develop nearly 200 well pads on the site, each covering approximately 5 acres. While the drop of oil and gas prices shortly afterward temporarily stalled exploration, this was a successful effort that created local jobs and helped to develop the region's rich shale gas deposits, and it still promises long-term financial gain for AEP. Over a 7-year period, AEP was paid an estimated $15 million for leasing rights, as well as royalties, which created income for AEP and its shareholders.

Fig. 3.9 Gas fracking rig and drilling pad at ReCreation Land

The relationship between AEP and the Columbus Zoo and Aquarium, an important stakeholder that owns and operates The Wilds, became strained during the development of an adjacent site for natural gas. The Wilds is a nonprofit park and conservation center located on land that was donated by AEP in Cumberland, Ohio, that supports rare and endangered species – such as fringed-eared oryx, African wild dogs, sika deer, and southern white rhinos – that live in natural, open-range habitats. The animals are valuable, and any threat to them, whether real or perceived, is a concern to the Zoo, environmental advocates, and the local community.

In a February 2012 *Toledo Blade* article [3], it was noted that AEP had leased mineral rights below The Wilds to Anadarko Petroleum as part of the agreement that gave the company access to 125,000 acres of AEP-owned mineral rights in eastern Ohio. That announcement generated some concern among stakeholders, including the Columbus Zoo and Aquarium. Some traditional gas wells exist on the property, but none that use hydraulic fracturing. Environmental groups argued that spills and poorly designed wells could potentially poison streams and groundwater and harm wildlife, despite industry assurances that the safety measures they take prevent this from happening.

One of the concerns at The Wilds was the effect that lights or noise generated by the drilling operation would have on the animals. Lights or noise generated by a 24-hour-a-day fracking operation would be of great concern, as the preserve is a breeding facility that needs a normal light sequence. AEP and Anadarko promised that precautions would be taken to protect the animals and visitors when any drilling started, but The Wilds and Zoo managers remained skeptical. Anadarko installed sound and light barriers around drill sites and in some cases reduced operating hours at rigs to accommodate the animals' needs. An Anadarko spokesperson was quoted as saying, "We certainly recognize the unique environment and considerations that must be made in that area" [3]. In addition, AEP retained the right to approve all well locations and "would not do anything that would be detrimental" to The Wilds [3].

While the controversy surrounding the well work on The Wilds property was ongoing, AEP began an ecosystem service modeling study to ensure that the well pads on the ReCreation Land property were placed in areas that minimized ecosystem service impacts. The Wilds and Zoo organizations were offered the opportunity to participate in the study as well, but they declined.

With help from the Electric Power Research Institute (EPRI), AEP launched an evaluation of ecosystem service impacts caused by placing well pads on different sites throughout the property. The project team used a tool called Integrated Valuation of Environmental Services and Tradeoffs (InVEST) to map the site and assess the impacts [30, 31]. The ecosystem services analysis was conducted on more than 200 potential well pad locations for shale gas production to evaluate potential impacts to biodiversity, carbon sequestration, nutrient and sediment retention, and pollination. The study found that the modeled ecosystem impacts were relatively small overall and that those impacts could be mitigated within 30–40 years through active restoration to allow vegetation regrowth near the well sites. EPRI conducted further analyses to identify well pad locations that would be least impactful, and AEP used that information to identify 25 well pad sites. By optimizing the locations of these particular well sites, AEP was able to minimize impacts to biodiversity, carbon storage, sediment retention, crop pollination, and nitrogen and phosphorus retention (Fig. 3.10).

Briefings on this effort were provided to The Nature Conservancy, Stanford University, World Wildlife Fund, and other groups. The results were shared with faculty at The Ohio State University, which was using the project as a case study. Unlike the situation with The Wilds, there were no stakeholder disputes about the many wells that were drilled and fracked on the ReCreation Land property, partly due to the InVEST modeling results. However, the well on The Wilds' property was never drilled due, in part, to concerns about the exotic animals.

3.5.4 Private Versus Public Property Ownership

AEP offers more than 80,000 acres of property for public recreational activities (i.e., hiking, hunting, fishing), and ReCreation Land makes up more than 60,000 of that acreage. The company's thinking is that, since these properties are no longer being used for AEP's business purposes, the adjacent communities should have the opportunity to reacquire them and resume farming, building houses, pursuing economic development opportunities, or preserving the acreage for recreational purposes. However, this position has also proved to be somewhat controversial, as hunters and fisherman, particularly those who do not own large tracts of land, value the open access to ReCreation Land. AEP holdings in southeast Ohio have gone on the auction block several times in recent years. In the fall of 2015, the company generated $3.5 million from selling 1966 acres in four southeastern counties.

Disposing of that property meant dividing it into 66 parcels of 5–175 acres each, with selling prices averaging $1100 to $3600 per acre. AEP could dispose of the ReCreation Land property in a similar way, and there is considerable pressure to do so [32]. AEP management strives to generate income for earnings, and selling the property is always a possibility because there is no interest in reviving coal mining efforts, the purpose for which the land was originally purchased.

On May 5, 2016, the head of the Ohio DNR's wildlife division indicated that the agency was considering purchasing part or all of ReCreation Land, but no transaction was conducted at that time [32]. Hunting, trapping, fishing, hiking, and camping are currently free to the public via a permit system, and loss of the land for these purposes would be a major disappointment to communities and residents in the region who have come to rely upon open access to the property.

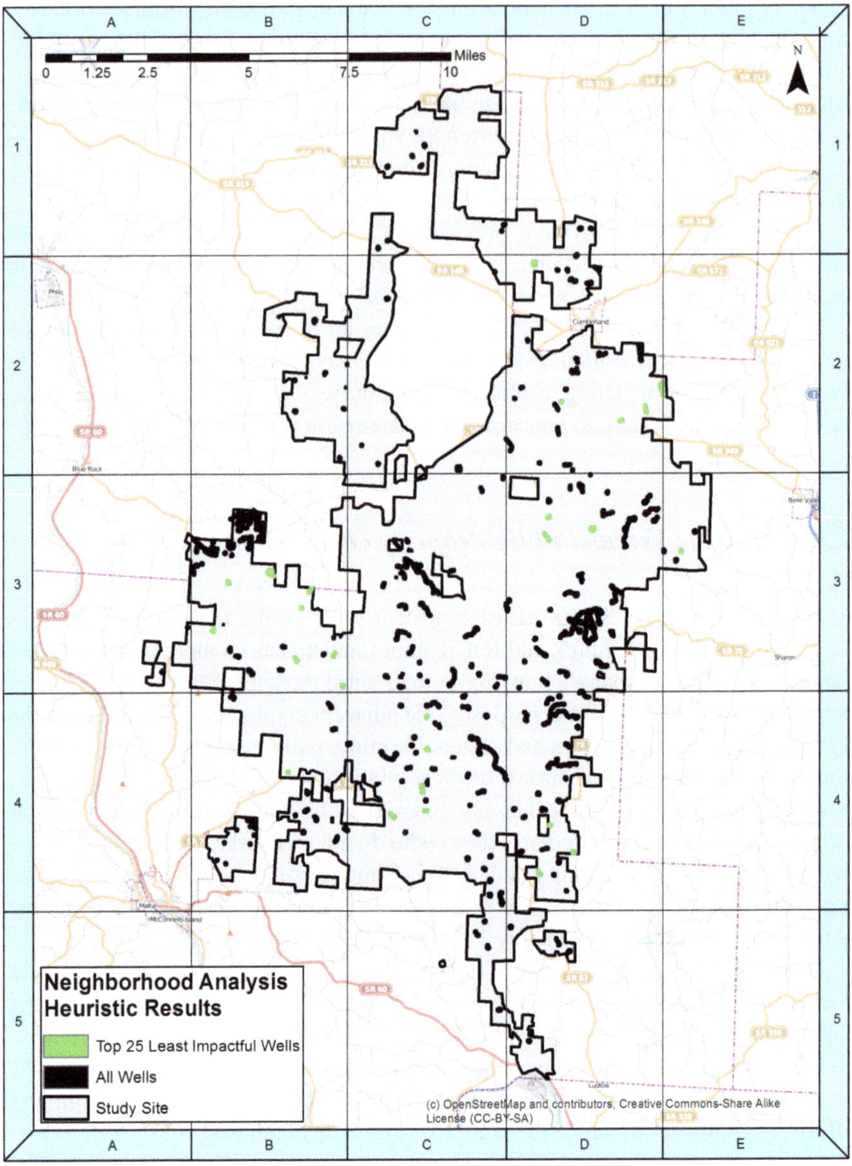

Fig. 3.10 Results of analysis to select 25 lower-impact well pad sites

3.5.5 Solar Development

Tremendous growth in renewable energy generation – solar energy, in particular – has occurred within AEP, and AEP business models indicate that as much as 8360 MW of wind and solar generation may need to be installed over the next

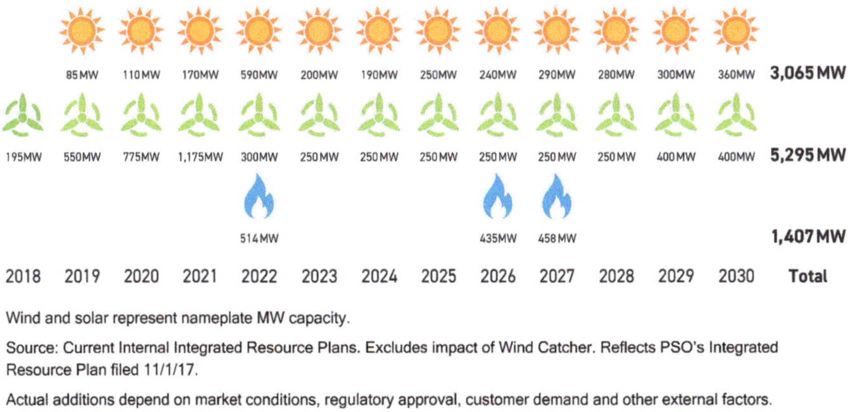

Fig. 3.11 Modeled AEP System generation resources additions

13 years (Fig. 3.11). AEP has been studying several blocks of acreage within ReCreation Land over the past 5 years to determine the feasibility of a renewable energy site, most likely solar. At present, there are two promising sites, with proximity to existing transmission lines and contour of ground being important criteria. The development of such a project would help the company meet its renewable energy goals and support the public's desire for "cleaner" sources of electricity.

While there are obvious benefits to solar generation, such as the low environmental footprint, no greenhouse gas emissions, no water or noise pollution, and an unlimited supply of power from the sun, there are concerns about environmental impacts. These include changes in land cover; water use and management issues, such as soil erosion and stormwater contamination; impacts to wildlife and habitat; and community impacts [33]. For example, birds and their insect prey can mistake the reflection of light from the solar panels to be a water body; as a result, they can suffer injury or death should they collide with the panels [34].

3.6 Future Opportunities

It has always been necessary to consider potential future uses of the ReCreation Land property, including more nontraditional options. A less obvious impact following timber, pasture, real estate, gas, or solar development is the future inability to establish "banks" for endangered species or specialized habitats such as wetlands. Regulations have been established to require mitigation or offsets for wetlands that have been impacted or destroyed during land development [35]. Developers can provide ecological conservation themselves or purchase offset credits from a "mitigation bank," which is a private enterprise that conserves large tracts of property for the purpose of selling the associated compensation "credits." AEP considered the use of ReCreation Land for the development of wetland mitigation banks in 2002 and could have realized a significant economic return from such a program [6]. However, concerns about restrictions on future development of the property prevented the implementation of such a program.

Similar to wetland banking, the US Fish and Wildlife Service has allowed the trading of endangered species habitats under the authority of the Endangered Species Act [36]. Under such an arrangement, a property owner who plans to take a land management action that may harm a federally listed endangered species is required to provide compensatory mitigation to improve conditions for that same species on a different property [37]. This could involve purchasing species conservation credits from a third-party mitigation bank that specializes in creating and restoring habitat for that particular species. Such a project could be carried out at ReCreation Land, where several rare, threatened, or endangered species live on or near ReCreation Land. Property could be set aside to provide habitat for a specific species in compensation for other property that is developed. However, if the property is sold, this opportunity is lost.

The ODNR has continued to express an interest in keeping the site preserved and open to the public, as it has been for more than 50 years. On August 9, 2017, the ODNR announced the signing of a memorandum of understanding (MOU) with AEP through which the agency will negotiate the purchase of a major portion of the ReCreation Land property [38]. This would achieve two objectives at once: generating desired income for AEP while allowing continued public access for recreational opportunities.

Few other organizations could afford to buy and maintain the land, and in the past, the challenge for the ODNR has been funding availability. Although the agency receives financial support from the US Fish and Wildlife Service, it would have to provide a 25% match to receive the additional funding from the federal agency needed to purchase the land. The financial match is usually derived from the sale of hunting and fishing licenses, but such income is not guaranteed and, in some cases, is declining [39].

Funding for the ODNR may still be an issue, and the MOU addresses only the intention of the two organizations to work together. Much remains to be done. Parcels have not been selected, and compensation has not been determined. Despite the outward appearance of a win-win being achieved, there would still be tradeoffs because AEP would forego the full commercial value of the property, while the state would have to assess taxes to maintain the land for public access.

3.7 Conclusion

In the book *Deciding What to Save: Trade-offs in Conservation* [40], the authors note that conservation actions inevitably involve choices: choices between native species and local ecosystems, between preservation and transformation by economic forces, between the needs of people and those of endangered species, and between the interests of some people over others. Very often, it is necessary to make tradeoffs, for example, by selecting some species and ecosystems for conservation while abandoning others or by taking one approach to conservation while ignoring others [40].

But do these tradeoffs matter? What further biodiversity is being lost because of them? Why should we conserve? What should be conserved? How much should be conserved, and how much should we pay to conserve it? There is often polarization over how to implement the goal of adopting economically and socially sound measures to incentivize conservation and sustainable use [40]. Conservationists want to preserve as much biodiversity as possible, but all conservation entails some form of cost for someone, somewhere.

These same lessons apply to the AEP ReCreation Lands. While AEP seeks to maximize its ability to produce revenue through gas leases, commercial sales, and other activities, this revenue may come at a cost to the public who use the site for recreation and depend on it for providing a healthy environment. At the same time, there is a cost even when preserving the site exclusively for ecological uses. The examples presented here illustrate how these costs can be balanced and compromises reached, but challenges remain.

So what will happen to AEP's ReCreation Land? Is there a balance that can be achieved to accommodate the future needs of the company, its shareholders, and the local community? John McManus, senior vice president of AEP Environmental Services, says that "in looking at these issues across our system, we try to reach a balance between the needs of local communities, the concerns of those who place a high priority on ecological protection, and the financial outcome for our customers and shareholders. Open and effective communication among interested stakeholders is required to reach this balance."[1]

The use of coal has decreased dramatically over the years, precluding any future mining. Revenue from gas fracking has also decreased, diminishing the need for new wells. Will AEP continue to sell portions of the site to the extent that its biological value is diminished, or will AEP invest in mitigation banking as a revenue source? Will portions of the site be developed for solar power as the demand for renewable energy grows, or will AEP reach an agreement with the ODNR to allow continued access to the property for recreation?

All of these options continue to be explored, but, as it has in the past, AEP will accommodate, as best it can, the needs of all its stakeholders. This is indeed a magical land and one of the largest open spaces in Ohio. But, unlike the fairytales of Narnia or Terabithia, the story of AEP's ReCreation Land is still being told, and the end has not yet been written.

Acknowledgements The authors acknowledge the support and assistance of Sandra Nessing and Melissa Tominack, who provided important guidance and help with the development of this chapter. Their insight, particularly as related to sustainability issues, proved invaluable.

[1] McManus, John. Personal communication. 15 December 2017.

Author Profiles

Timothy W. Lohner Dr. Lohner is a consulting environmental specialist at American Electric Power (AEP). Prior to joining AEP, Dr. Lohner was employed by Virginia Power, the Ohio Environmental Protection Agency, and the Cincinnati Gas & Electric Company. At AEP, he addresses issues involving the implementation of water quality standards, the permitting of carbon capture and storage projects, the development of state and federal regulations governing water resources, environmental disclosure reporting, vegetation management for biodiversity, and the implementation of ecosystem resource projects in the Ohio River basin. Dr. Lohner holds a bachelor's degree in biology from the University of Cincinnati and masters and doctorate degrees in environmental toxicology from the Ohio State University. He currently serves as an advisor to the Electric Power Research Institute, the Utility Water Act Group, the Edison Electric Institute, and the Columbus State Environmental, Health and Safety Program. He has received multiple research awards and has published numerous articles on aquatic toxicology, risk assessment, and environmental policy and is co-author of the book *Sound Science, Junk Policy*.

C. Michael Williams Williams earned bachelor's degrees in education and photography from Ohio University in 1991. He began working with AEP in 2000 as a geographic information systems (GIS) coordinator in the Forestry Department, assisting the group with their effort to sustainably manage more than 160,000 acres of standing timber.

Williams is currently in a real estate supervisor role with AEP. Alongside a staff of 30 people, he assists in the management of over 350,000 acres of ground in 11 states. His responsibilities include acquiring property for generation and transmission use, sale of excess ground, and oversight of the Forestry Recreation and GIS sections of the department. He works with state and federal agencies to manage over 85,000 acres for public recreation – including hiking, fishing, birdwatching, and hunting – within AEP ownership.

Williams and his wife Renee live in southeast Ohio's Hocking Hills region, just south of ReCreation Land.

Jeffrey W. Wilson Wilson earned a bachelor's degree in business from Glenville State College and an associate's degree in forestry from Hocking College. Wilson began his career in 1996 as a forester working on the climate change initiative by overseeing the planting of three million trees per year and completing selective forest management sales to sequester greenhouse gases.

Wilson is currently a forester Sr. with AEP in the Forestry Recreation and GIS sections of the Real Estate Asset Management Group. He is involved with many aspects of the management of AEP real estate, including the management of approximately 160,000 forest acres, sales of excess real estate, leasing of company assets, GIS mapping, and recreational program oversight. He has been trained by the forest service in silviculture, is an instructor at Ohio Forestry and Wildlife camp, and is a certified tree farm inspector.

Along with his family, he manages an 800-acre farm in southeast Ohio adjacent to ReCreation Land. He and his wife Carla have three children and reside on the farm.

Company Profile

Headquartered in Columbus, Ohio, American Electric Power (AEP) is one of the largest electric utilities in the USA, serving nearly 5.4 million customers in 11 states. Its service territory covers 200,000 square miles in Arkansas, Indiana, Kentucky, Louisiana, Michigan, Ohio, Oklahoma, Tennessee, Texas, Virginia, and West Virginia. AEP owns nearly 26,000 megawatts of generating capacity in the USA. AEP also owns the nation's largest electricity transmission system, a more than 40,000-mile network that includes more 765-kilovolt extra-high-voltage transmission lines than all other US transmission systems combined. AEP's generation and marketing competitive business segment includes subsidiaries that have non-utility generating assets, a wholesale energy trading and marketing business, and a retail supply and energy management unit.

References

1. Lewis, C.S. *The Lion, the Witch and the Wardrobe*. Geoffrey Bles, 1950.
2. Paterson, Katherine. *Bridge to Terabithia*. Harper Collins, 1977.
3. Blade News Services. "Mineral Rights Deal Would Allow Natural Gas Shale Drilling at Ohio Exotic Animal Preserve." *Toledo Blade*, 29 February 2012. www.toledoblade.com/State/2012/02/29/Mineral-rights-deal-would-allow-natural-gas-shale-drilling-at-Ohio-exotic-animal-preserve.html. Accessed 4 February 2018.
4. Hunt, Spencer. "Shale Wildcatters May Join Wilds Menagerie." *Columbus Dispatch*, 29 February 2012. www.dispatch.com/article/20120229/NEWS/302299645. Accessed 4 February 2018.
5. *Data USA: Morgan Co. Ohio*. https://datausa.io/profile/geo/morgan-county-oh/#economy. Accessed 4 February 2018.
6. *Ecological Asset Value Development Report American Electric Power, ReCreation Lands, OH: Phase I and Phase II*, EPRIsolutions, Environment Division, Palo Alto, CA. 2002.
7. United States, Fish and Wildlife Service, IPaC, *Information for Planning and Consultation*. Version 1.4, 2015; http://www.wind.tnc.org/#. Accessed 4 February 2018.
8. Sneed, Annie. "The Search is On for Pulling Carbon From the Air." *Scientific American*, 27 December 2016. www.scientificamerican.com/article/the-search-is-on-for-pulling-carbon-from-the-air/. Accessed 4 February 2018.
9. Gallai, Nicola, et al. "Economic Valuation of the Vulnerability of World Agriculture Confronted with Pollinator Decline." *Ecol. Econ.* Vol. 68, 2009, pp. 819–821.
10. Klein, Alexandra-Maria, et al. "Importance of Pollinators in Changing Landscapes for World Crops." *Proc. R. Soc.* Vol. 274, 2007, pp. 303–313.
11. Morse, Roger and Nicholas Calderone. "The Value of Honey Bees as Pollinators of U.S. Crops in 2000." *Bee Culture*, March 2000, pp. 1–15.
12. Boyles, Justin, et al. "Economic Importance of Bats in Agriculture." *Science*. Vol. 332, 2011, pp. 41–42.

13. Harris, Ryan. “Reshaping the Land for Wildlife.” *Wild Ohio*, January/February 2015. pp. 20–21.
14. Duren, Ken. “Coal Mine to Gold Mine.” *Wild Ohio*, March/April 2015, pp. 22–23.
15. Galicia, Leopoldo and Esmeralda Zarco-Arista. “Multiple Ecosystem Services, Possible Trade-offs and Synergies in a Temperate Forest Ecosystem in Mexico: a Review.” *Intl. J. Biodivers. Sci. Ecosystm. Serv. & Manage*. vol. 10, no. 4, 2014, pp. 275–288, https://doi.org/10.1080/21513732.2014.973907. Accessed 16 February 2017.
16. Vihervaara, Petteri, et al. “Biodiversity Trade-offs and Globalizing Forestry.” *Forest Policy and Economics*. Vol. 26, 2013, pp. 147–148.
17. Kuehler, Natalie. “Why Thinning Forests Can Be Good.” *National Forest Foundation*. 30 April 2015. www.nationalforests.org/blog/why-thinning-helps-the-forest. Accessed 4 February 2018.
18. Daymond, John, et al. “Tradeoffs Between Soil, Water and Carbon – a National Scale Analysis from New Zealand.” *J. Environ. Manage*. Vol. 95, 2012, pp. 124–131.
19. American Tree Farm System. “2010–2015 Standards.” ATFS, May 2010. www.treefarmsystem.org/certification-american-tree-farm-system Accessed 4 February 2018.
20. United States, Environmental Protection Agency. *Protecting Water Quality from Agricultural Runoff*. EPA–841–F–05–001. 2005. *www.epa.gov/sites/production/files/2015-09/.../ag_runoff_fact_sheet.pdf*. Accessed 4 February 2018.
21. United States, Environmental Protection Agency Memorandum. *Nutrient Pollution and Numeric Water Quality Standards*. 2007. www.epa.gov/nutrient-policy-data/memo-nutrient-pollution-and-numeric-water-quality-standards. Accessed 4 February 2018.
22. Wines, Michael. “Toxic Algae Outbreak Overwhelms a Polluted Ohio River.” *New York Times*, 30 Sept. 2015. https://www.nytimes.com/2015/10/01/us/toxic-algae-outbreak-overwhelms-a-polluted-ohio-river.html Accessed 4 February 2018.
23. Laubel, A., et al. “Bank Erosion in a Danish Lowland Stream.” *Hydrobiologia*. Vol. 410, 1999, pp. 279–285.
24. McDowell, Richard and Robert Wilcock. “Sources of Sediment and Phosphorus in Stream Flow of a Highly Productive Dairy Farmed Catchment.” *Journal of Environ. Qual*. Vol. 36, 2007, pp. 540–548.
25. Davies-Colley, Robert. “Stream Channels are Narrower in Pasture than in Forest.” *New Zealand. J. of Marine and Freshwater Research*. Vol. 31, 1997, pp. 599–608.
26. Ohio, Department of Natural Resources, Mineral Resources Management. *Ohio Hydraulic Fracturing State Review*. ODNR, 2011.
27. Burnham, Andrew, et al., “Life-Cycle Greenhouse Gas Emissions of Shale Gas, Natural Gas, Coal, and Petroleum.” *Environ. Sci. Technol*. Vol. 46, No. 2, 2012, pp. 619–627.
28. National Research Council. *Induced Seismicity Potential in Energy Technologies*. Washington, DC: The National Academies Press. 2013. https://doi.org/10.17226/13355 Accessed 4 February 2018.
29. *United States, Environmental Protection Agency*. Providing Regulatory Clarity and Protections against Known Risks. https://www.epa.gov/hydraulicfracturing#providing. Accessed 4 February 2018.
30. Keller, Arturo, et al. “Minimizing Impacts of Land Use Change on Ecosystem Services Using Multi-Criteria Heuristic Analysis.” *J. Environmental Mgmt*. Vol. 156, 2015, pp. 23–30.
31. Electric Power Research Institute. *Assessing Ecosystem Services: Using the InVEST Model: Case Study of the American Electric Power ReCreation Lands, Ohio*. EPRI, Palo Alto, CA. 3002005275. 2015.
32. Beathard, Jane. “State in Talks Over Fate of AEP ReCreation Land.” *Outdoor News*, 5 May 2016. http://www.outdoornews.com/2016/05/05/state-in-talks-over-fate-of-aep-recreation-land/ Accessed 4 February 2018.
33. Electric Power Research Institute. *Environmental Aspects of Solar Interest Group*. EPRI, Palo Alto, CA. 2016.

34. Upton, John. "Solar Farms Threaten Birds." *Scientific American*. 27 August 2014. www.scientificamerican.com/article/solar-farms-threaten-birds/. Accessed 4 February 2018.
35. U.S. National Archives and Records Administration. *Code of Federal Regulations*. CFR 40 Part 230 Section 404(b)(1). Specification of Disposal Sites for Dredged or Fill Material. 1995.
36. Fox, Jessica and Anamaria Nino-Murcia. *"Status of Species Conservation Banking in the United States."* Conservation Biology. Vol. 19 No. 4, 2005, pp. 996–1007.
37. Hanley, Nick, et al. "How Should We Incentivize Private Landowners to 'Produce' More Biodiversity?" *Oxford Review of Economic Policy*. Col. 28 No. 1, 2012, pp. 93–113.
38. ODNR.gov. ODNR in Discussions with AEP for Potential Expansion of Outdoor Recreation Opportunities. *Ohio Department of Natural Resources. 9* Aug. 2017. http://ohiodnr.gov/news/post/odnr-in-discussions-with-aep-for-potential-expansion-of-outdoor-recreation-opportunities. Accessed 4 February 2018.
39. Beathard, Jane. *Outdoor News*, 5 May 2016.
40. Leader-Williams, Nigel, et al. "Deciding What to Save: Trade-offs in Conservation." *Trade-offs in Conservation*, edited by Nigel Leader-Williams, et al., Blackwell Publishing, Ltd., 2010, pp. 3–13.

Chapter 4
Sustainable Electric Power from a Responsible Investing Perspective

Markus Walther

Abstract Investors can play an important role in making sustainability considerations a strategic issue for corporate executives. The increasing availability of corporate sustainability data has led to investment approaches that consider sustainability metrics in investment processes. Investors incorporating environmental, social, and corporate governance (ESG) criteria to generate long-term competitive financial returns and positive societal impact – so-called socially responsible investors (SRI) or ESG investors – have experienced significant client growth. This chapter describes ESG investing and explores the extent to which financial performance and sustainability are compatible and even mutually reinforcing. The concept of materiality is discussed, illustrating the need to focus sustainability efforts on areas that have financial importance to a company. A perspective on the electric power industry through the lens of an ESG investor is offered along with a discussion of investor challenges posed by certain trends in the industry, such as load defection, stranded asset risk, and climate change. While financial performance and sustainability are compatible in many instances, there are also a series of tradeoffs ESG investors face when financial or business needs collide with sustainability issues or one area of sustainability competes with another.

4.1 Introduction

One afternoon in the summer of 2017, Calvert staff had a lively discussion on the electric industry. Calvert's CEO, members of the equity group, the fixed income group, and the environment, social, and governance (ESG) research department sat in a room discussing our approach to investments in the electric power industry. Generally, we were in agreement on which utility companies we considered leaders in sustainability and potentially attractive investments and which utilities would definitely not meet our principles of responsible investing. However, there were many companies somewhere in the middle for which we needed to come up with a

M. Walther (✉)
Independent, Washington, DC, USA

J. Fox, M. Scott (eds.), *Sustainable Electricity II*,
https://doi.org/10.1007/978-3-319-95696-1_4

decision. Two of these companies in particular were the subject of an animated debate.

The two companies in question both had large fleets of coal power plants, which raised some eyebrows. At the same time, both also had some traits that we liked; one of the two companies had made significant investments in grid modernization technologies like electric storage and was involved in providing electricity in developing countries, while the other had strong environmental practices and maintains commitments to reduce carbon emissions.

Should a firm like Calvert avoid investments in companies with potentially significant stranded asset risks[1]? When do such risks become significant? Does the answer change if the company is undertaking serious efforts to become more sustainable? What would Calvert's clients think? And how should we compare financial considerations with societal aspects, like providing electricity in developing countries or reducing air emissions and water use?

These are questions that socially responsible investors like Calvert frequently have to tackle when considering investments in the electric power industry. They are at the core of the pursuit to make investments that are financially lucrative and societally beneficial, or to balance the two when tradeoffs are necessary.

This chapter first provides background on mutual funds, ESG investing, and how recent trends have facilitated the growth of funds combining classic financial metrics and sustainability indicators in their investment processes. The second part of the chapter describes why the electric power industry is a challenging area for ESG investors in general and what particular tradeoffs ESG investors face in the power industry. The second part also provides four practical examples of companies, illustrating to what extent sustainability and financial performance can be combined.

4.1.1 Investing Basics

Utilities finance themselves by issuing stocks and bonds in roughly equal parts. The stocks and bonds are bought by investors such as banks, insurance companies and private equity firms, as well as asset managers like hedge funds, exchange-traded funds, and mutual funds. Investors in all these categories try to achieve high returns while taking on limited risk, but they vary in return goals, risk appetite, and investment horizon. While many people may not spend much time thinking about these nuances, anyone with a retirement plan is at least indirectly an investment client of a mutual fund or exchange-traded fund.

[1] Stranded asset risk is the possibility that an asset, like a power plant, may become obsolete before the end of its expected life and therefore generate less value than its owners or investors had expected.

Mutual funds invest on behalf of clients such as pension funds, foundations, and private individuals. They pool client money into funds that own a portfolio of stocks and bonds – perhaps 50-100 per fund – with the implication that the mutual fund will achieve better financial performance than the client could. Achieving better financial performance means providing strong returns with limited risk. This can be accomplished by finding investments that offer returns at least commensurate with their risk and which ideally are undervalued by other investors. For instance, noticing a little-known company that has better growth prospects than most investors realize, or a struggling company that is not in as much trouble as others think, can yield a higher return than what the company's risk profile would warrant. Note a key implication: an investor who has better tools to spot risks and opportunities than other investors can deliver better financial performance.

The other aspect of financial performance is diversification. Whereas a private person might buy stocks of a handful of large, well-known American companies, mutual funds create portfolios with dozens of companies of various sizes, industries, and geographies (Figs. 4.1 and 4.2). The idea is to create hedges so that if, say, a surge in copper prices hurts the profits of Korean electronics manufacturers held in a fund, their value decline is balanced by the rising stock of European mining conglomerates. Funds typically also benchmark their performance against an established index like the S&P 500, which means they cannot deviate too far from its composition and therefore have an incentive to invest in most industries.

Another common category of investors are exchange-traded funds or ETFs. While exchange-traded funds are a fairly new class of investors, they have attracted a lot of money in recent years due to their low fees. ETFs are similar to mutual funds

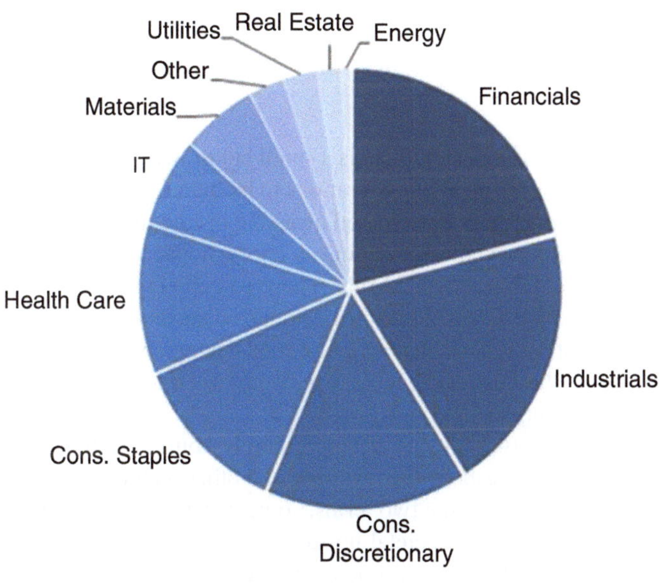

Fig. 4.1 Sector allocation of Calvert Equity Fund [1]

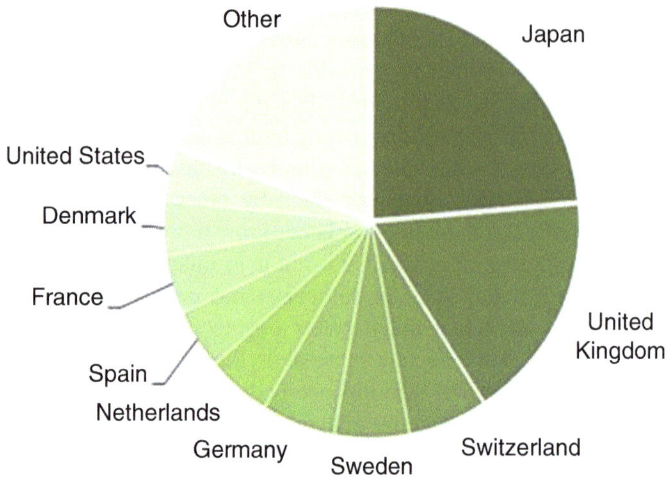

Fig. 4.2 Geographic allocation of Calvert Equity Fund [1]

in that they pool client money and invest through diversified portfolios. However, two key differences are that ETFs generally do not conduct in-depth research on companies and they aim for average returns rather than trying to beat the market; their portfolios typically follow an index. For example, if Apple accounts for 2% of the MSCI World Index and Energias de Portugal for 0.02%, an ETF tracking the MSCI World Index would automatically invest 2% of its funds in Apple and 0.02% in Energias de Portugal.

4.1.2 ESG Investors

Within the mutual funds industry, there is a small but growing segment of firms that have expanded their investment considerations to include ESG criteria. Historically, the first examples were funds that avoided investing in certain companies for ethical reasons, such as faith-based funds avoiding alcohol or tobacco or Calvert not investing in South Africa during apartheid rule. These so-called negative screens can be contrasted with positive screens, meaning a focus on societal benefits such as promoting human rights or renewable energy. A good example would be impact investors, for whom tangible societal benefits may be just as important, or even outweigh strong investment returns.

In recent years, the two screens are increasingly combined under an approach some refer to as socially responsible investing and others call ESG investing. There are various interpretations of the two terms, some considering socially responsible investing as more focused on societal impact; for simplicity's sake, the term ESG investing is used and means an investment approach that incorporates various ESG metrics for two purposes: to benefit society and to improve financial performance.

For the first part, benefitting society, Calvert considers the impact companies have on society and the environment. Preference is given to companies that take care of their employees, offer products and services that are safe for consumers and the environment, and promote transparent business practices. Conversely, a company will be considered less sustainable if it exacerbates social or environmental problems or engages in unethical behavior. Examples would be textile companies indifferent to child labor in their supply chain, electronics manufacturers not phasing out toxic components, or deceptive sales tactics and political maneuvers of some solar developers [2] and electric utilities [3].

For the second part, improving financial performance, the underlying idea is that some ESG metrics may provide clues to a company's future risks and opportunities and so-called "material" issues should therefore be included in investment research. What are "material" ESG issues? Let us look at some examples.

> Investors understand that companies that make more efficient use of resources, reduce emissions, eliminate or reduce waste streams and make progress on the spectrum of sustainability opportunities, are usually improving the economics of the business, or reducing risk. The metrics around these efforts and the linkage to the company's economics are of great interest to investors.
> – John Streur, CEO, Calvert Research & Management.[2]

For corporate governance, one aspect is the effort dedicated to prevent corruption. Alstom paid bribes to secure contracts for over a decade, until its endemic corruption prompted investigations in various countries and resulted in a $772 million fine in 2014. For social metrics, one example is how well a company protects customer data. When consumer credit reporting firm Equifax got hacked in September 2017, its stock price tumbled by a staggering 35% within 7 days and was far from a full recovery as of the writing of this book. For environmental metrics, one factor is the focus of a company on developing more sustainable products. Truck engine manufacturers Cummins and Navistar took opposite approaches in 2008; Cummins invested in more efficient engine designs, and its products sold well. Navistar in contrast refrained from adopting new technology designed to minimize emissions and took shortcuts in product testing. Navistar assured customers and investors that its engines complied with tightening emission standards, when in fact they did not. Ensuing fines and lawsuits cost the company millions, and its stock underperformed for 2 years.

For any of these examples, financial metrics at the time likely would not have indicated trouble down the road, yet all these ESG issues clearly had a direct impact on shareholder returns. One challenge for ESG investors is to know which ESG issues to focus on for a given company. As with financial metrics, this is currently still more art than science. The other question is where to get the right ESG data.

[2] Personal communication. October 28, 2017.

Here, increasing corporate disclosure on sustainability themes has dramatically changed the toolkit of ESG investors in recent years.

Pioneered by European firms, the practice of publishing annual sustainability reports is spreading across the globe. While corporate disclosure efforts vary significantly in scope and quality across regions and industries, there is increasing transparency into how companies manage sustainability issues. There are also initiatives to standardize such disclosure, such as the Global Reporting Initiative and the Sustainability Accounting Standards Board. As a result, data vendor companies gather and aggregate sustainability metrics and sell them to ESG investors. It is conceivable that at some point in the future, sustainability disclosure will go the same way as financial disclosure and simply become a standardized practice that investors expect.

ESG investing has grown significantly in recent years [4], driven by increasing demand. As the share of women and millennials grows among clients, financial advisors report increasing interest in ESG investing [5]. Both institutional clients such as pension funds as well as individuals increasingly want to see their invested money have a positive impact in the world. Whereas traditionally many clients and investors assumed a tradeoff between sustainability and financial performance, a growing body of research examining the relationship between ESG investing and financial performance has found that incorporating broader sustainability considerations into the investment process does not necessarily reduce financial performance and may actually enhance it [6].

To summarize, ESG investors invest client money into portfolios of companies selected according to financial and ESG metrics with the goal of achieving strong financial performance and positive societal impact for the client.

4.2 Sustainable Investing in the Electric Power Industry

Let us look at the power industry from the perspective of an ESG investor. I will first provide a high-level perspective, summarize some sustainability tradeoffs ESG funds face in the power industry, and lastly provide some practical investment examples.

4.2.1 Electric Power: A Challenging Industry for ESG Investors

The electric power industry is part of the utility sector. Utilities and energy together account for about 10% of the market value of global stocks, and by extension, utility and energy companies account for a sizable portion of established indices like the S&P 500. Therefore, an ESG investor tracking such indices would theoretically want to allocate about 10% of its funds to companies in these two sectors. However, the energy sector consists mostly of oil and gas companies, which are often not what sustainability-focused clients have in mind in terms of societally beneficial impact.

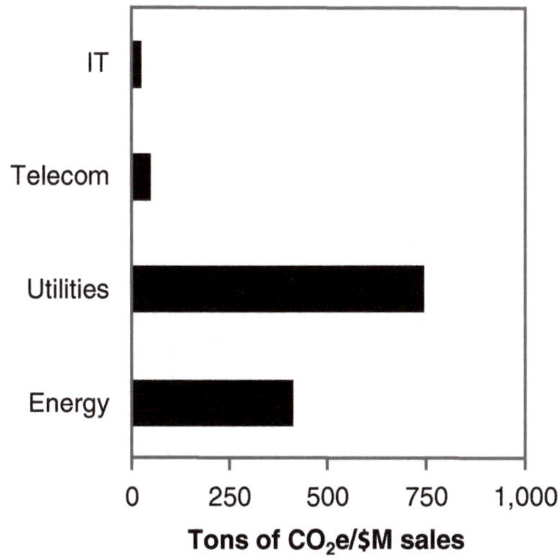

Fig. 4.3 Median carbon intensity for select sectors. (Based on analysis of Calvert data sourced from MSCI)

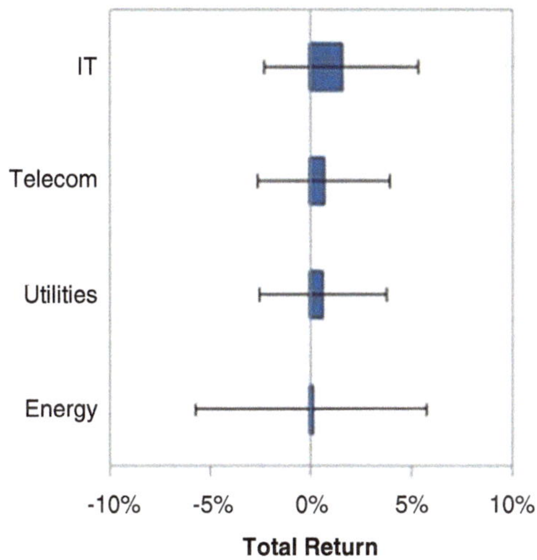

Fig. 4.4 Average monthly return and volatility for select sectors, 2011–2017. (Based on analysis of Calvert data sourced from MSCI)

ESG investors therefore look to the utility sector, which is dominated by the electric power industry. This industry comes with its own set of challenges, however.

The obvious challenge stems from clients' societal impact expectations regarding climate change. While we owe much of humanity's prosperity gains over the last hundred years to cheap and reliable electricity, it is also a fact that electricity generation accounts for about 25% of global greenhouse gas emissions (Fig. 4.3) [7]. Many clients expect ESG firms to invest in a climate-friendly fashion.

Perhaps less obvious are the financial performance implications of the ongoing transition in the power industry, so it is worth exploring this aspect in more detail. Let us first consider why funds invest in electric utilities. Utilities typically cannot offer much growth; their current appeal to investors lies in steady returns at the very low risk of a regulated monopoly (Fig. 4.4). However, both returns and risks may be unfavorably affected by the longer-term trends of the industry. Regarding returns, utilities seem likely to become less attractive; electricity demand is stagnating or even shrinking, making continued revenue growth more challenging. Regarding risk, regulated electric utilities will for the most part remain low-risk investments in the near term. So far, bankruptcies are rare, and credit ratings are solid. However, when looking into the future, some questions appear.

Let us remind ourselves that compared to other industries, the electric power industry is extremely capital-intense. For example, a utility wanting to build a natural gas combined-cycle power plant with a capacity of 700 megawatts would need to raise capital of more than $700 million dollars [8]. To recoup that investment, that plant would then have to operate for decades. In other words, the value of a utility stock depends on stable long-term revenues. In theory, if a public utility commission approves building a plant, it guarantees that the utility company will recover its investment. However, some investors think that in the long term, this guarantee could become politically untenable in some states if it means forcing customers to pay for service they can get cheaper and more tailored to their needs elsewhere. If that possibility sounds far-fetched, let us look at some trends.

In the United States, the total cost of residential solar photovoltaic systems has dropped by about 61% between 2010 and 2017 [9], leading to rapid growth in residential installations. By 2015, the annual addition of residential solar capacity exceeded 2000 megawatts (Fig. 4.5) [10]. In California, residential customers generated an estimated 5.4% of their needed electricity with their own solar panels in 2016 [11]. As the cost of energy storage systems decreases, it remains only a matter of time until it becomes feasible for most residential customers to generate their own power and use their utility company just as a backup. A corresponding study by the Rocky Mountain Institute concludes this option is already economically viable in Hawaii and will become viable across the United States by the 2040s [12].

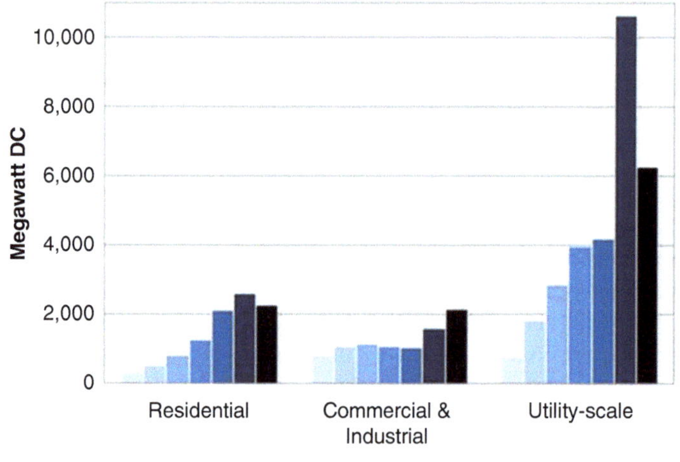

Fig. 4.5 US annual capacity addition in solar PV, 2011–2017 [13]

That trend of so-called load defection[3] is even more pronounced among commercial customers. Two factors are driving this trend. One is economics; wind is often the cheapest source of power, and solar energy is increasingly competitive as well. The other is climate change. Large corporations increasingly take action to reduce their carbon footprint, motivated by a mixture of corporate responsibility and a desire to protect their brand among consumers and employees. As of 2016, 211 of the Fortune 500 companies had targets to reduce their carbon footprint, and 53 of Fortune 500 companies had renewable energy targets [14]. Renewable energy contracts signed by corporations were equivalent to about 10% of new utility-scale wind and solar capacity in 2016 and reached at least 3100 megawatts in 2017 [15, 16]. These contracts are typically made with independent power providers, not utilities. MGM Resorts, the casino operator that defected entirely from its utility to contract with independent suppliers, is an outlier but also a warning to investors that electric utilities are losing their firm hold on formerly captive customers.

Besides the risk of losing customers, utilities also face stranded asset risk. In recent years, many coal power plants had to be written off because low gas prices made them unprofitable [17]. Over the long term, climate policy poses additional stranded asset risks. Granted, there is often a significant lag until public policy heeds the warnings of scientists and takes appropriate action, and climate change is no different. However, the trend seems clear also in the United States. Many states are gradually strengthening renewable energy mandates; the Supreme Court ruled that greenhouse gases are to be regulated as pollutants; and over the long term, despite the fossil fuel-friendly stance of the current administration, further attempts at curbing greenhouse gas emissions through federal regulations seem all but certain.

What is more, nobody can guarantee that change will come only gradually; utilities and their investors need to consider the possibility that concerns over the climate could lead to a turning point in climate policy. For an illustration, consider how in Germany long-standing concerns – in this case, not about carbon but the safety of nuclear energy – suddenly translated into a mandated phaseout of nuclear power plants when the Fukushima nuclear disaster struck in 2011.[4] In the United States, a similar reaction against fossil power plants seems unlikely at least at the federal level, but even if such a shift took place over a 10-year span, utilities could potentially lose assets worth billions of dollars. Funds need to take such risks into account long before they become likely, because once such risks become a concern for the broader investor community, stock values could already decline long before the risks actually materialize.

To summarize, electric power is a challenging industry for ESG firms. Many companies in it expose funds to reputational risks, and electric utilities are facing trends that gradually challenge the notion that they are virtually risk-free invest-

[3] Utility customers meeting part of their electricity demand with their own power generation, typically solar PV

[4] For a summary of how Germany's nuclear phaseout came about, see Ref. [33].

ments. In addition, ESG firms have to manage a series of tradeoffs in this industry, as we will see in the next part of this chapter.

4.2.2 Sustainability Tradeoffs

In theory, Calvert has simple criteria to find suitable investments in the power industry: companies that contribute strong financial performance and manage their societal impact well. For electric utilities, that would mean steady dividends and a rising stock price in line with peers, as well as good ESG performance. The latter concerns particularly environmental areas like air emissions, water use, and safe handling of coal ash waste but also includes an effective board of directors and a solid safety record. In reality, however, there are a number of tradeoffs that need to be balanced. They include practical business needs as well as competing sustainability interests.

In terms of tradeoffs related to business needs, the obvious first case is when environmental or social sustainability conflicts with financial performance. Many renewable energy funds, including a dedicated Calvert fund, saw terrible returns in 2010 and 2011 (Fig. 4.6), even if such investments may have helped the industry reach the scale necessary to become cost competitive. Conversely, the gas pipeline industry currently looks financially appealing, prompting some electric utilities to expand into this business. Sustainability analysts are ambivalent about expanding gas assets as I will discuss further below, whereas colleagues in the investment departments appreciate the growth potential. An important factor is that most mutual funds, ESG or not, have a short investment horizon; they hold stocks of a given company only for perhaps 1–3 years and then sell again when the stock of another, similar company is priced more favorably.

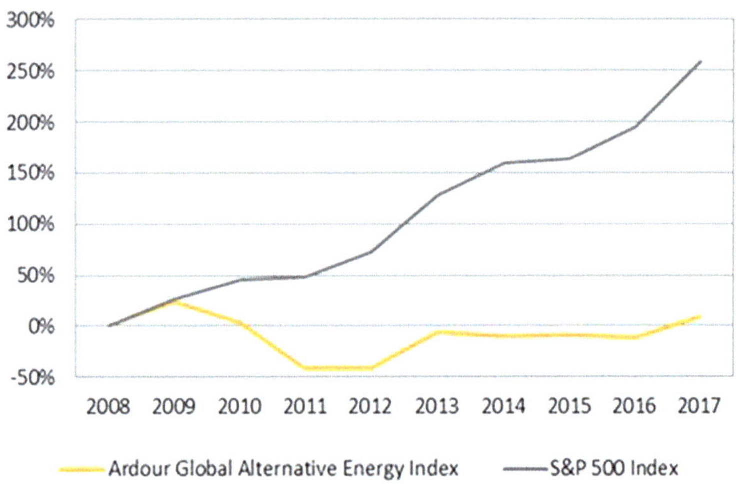

Fig. 4.6 Total return of sample alternative energy index [18]

A second tradeoff comes from the marketing needs of ESG firms. Sustainability is an important sales factor for ESG funds. Given that traditional mutual funds cannot compete with the low fees of exchange-traded funds, offering sustainable investments is a differentiating factor to justify higher fees. With the growth of ESG investing, clients can choose from an increasing number of funds, and ESG firms have to think about their image. This second tradeoff is worth expanding on especially regarding the power industry. The movement to divest from fossil fuels appears to have gathered momentum. The so-called divestment campaign started with university endowments, now numbering a few dozen,[5] divesting from fossil fuel-related companies. The direct financial impact is likely negligible, but that is not the point – we may be witnessing the rise of a stigma associated with fossil investments.

Noteworthy investors considering or implementing divestment initiatives include several state pension funds[6] and even Norway's $1 trillion sovereign wealth fund [19]. The insurance giant AXA announced its first investment restrictions around coal in 2015 and doubled down in December 2017 with the statement that it would stop investing in companies deriving more than 30% of their generation from coal [20]. JP Morgan, not generally considered a tree-hugger investor, said in 2016 that it will no longer finance new coal-fired power plants in high-income countries [21]. Similarly, ING announced that after 2025, the bank will no longer finance utilities that rely on coal for more than 5% of generation [22]. Meanwhile, the World Bank announced that it will no longer finance upstream oil and gas after 2019 [23]. An ESG investor may feel pressure to avoid heavy emitters or provide a good justification to clients.

A third tradeoff is the role of shareholder advocacy. Calvert and other firms regularly approach companies and request disclosure of sustainability metrics or adoption of certain sustainable practices. Calvert sees these dialogues as a way of nudging companies to improve their ESG profile, thereby hopefully improving their appeal to investors as well as their societal impact. However, these benefits may take a long time to materialize, and shareholder advocacy is labor-intensive, a challenge when mutual funds face increasing pressure to reduce their fees.

Cost pressure also drives a fourth tradeoff in the ESG investment industry: algorithms. In the past, a sustainability analyst would focus on a small set of companies and conduct in-depth research of their ESG profile. That is not a cost-effective way to compete with cheap exchange-traded funds, so ESG firms like Calvert use scoring models. Data vendors supply a host of ESG metrics, such as carbon emissions or quality of workplace safety programs, which models convert into ESG scores used to inform investment decisions. While the growing availability of sustainability data on companies provides helpful context, it is still often not sufficiently pre-

[5] For a list of institutions participating in fossil fuel divestment, see www.gofossilfree.org/divestment/commitments

[6] For instance, as of the writing of this book, four large state pension funds in California and New York have either stopped investing in coal companies and/or are debating fossil fuel divestment initiatives.

cise, recent, or meaningful enough to serve as a base for investment decisions. Consequently, if ESG funds tout their sustainability credentials based on quantitative methods, their marketing practices may merit a sustainability scrutiny of their own.

Besides these tradeoffs between sustainability and business needs, ESG investors also face tradeoffs between competing sustainability issues. Some of these conflicts relate to various impacts of fuel choices and the speed of emission reductions. One such tradeoff is natural gas. Fracking has reduced the price of gas so much that the shift from coal to gas plants has been the dominant driver in reducing carbon emissions from the US electric grid. Fast-ramping gas plants can also facilitate the large-scale integration of wind and solar energy. At the same time, gas cannot deliver the drastic additional reductions needed to avert the worst of climate change and does not look like a long term solution.

Another example of competing sustainability issues is the construction of hydroelectric dams in emerging markets or developing countries. At first glance, they seem the perfect case of the triple bottom line: emissions-free renewable power that supports the economic development of countries like Brazil, Turkey, or Ethiopia. However, such dams are periodic headaches for ESG analysts. There are several examples of dams being built in blatant violation of the rights of people living in the area, who often happen to be marginalized groups, like indigenous tribes in Brazil [24] or the Kurds in Turkey [25]. Evidently, large infrastructure projects rarely go without controversy, but client letters often express the view that dams resulting in the forced displacement of thousands are not sustainable, regardless of the eventual carbon benefits.

Finally, sustainability tradeoffs are also evident in the delicate issue of nuclear energy. The climate benefits are undeniable, and one could argue that in countries with strong regulatory oversight, the safety risks of this technology are worth taking to prevent the larger climate risks of replacing a significant share of generation with gas or even coal power. However, that is definitely not the view many Calvert clients have, and Calvert avoids utilities depending heavily on nuclear power.

These tradeoffs illustrate that one cannot always prevent hard choices between financial and other practical needs on one side and various sustainability needs such as safety, economic development, or low carbon emissions. ESG investors have to prioritize.

4.2.3 Practical Examples of Financial and Sustainability Tradeoffs

> Companies are demanding access to clean energy and investors are pressing companies to source 100% renewable. It is essential that utilities develop a comprehensive strategy to reduce climate impact, and clearly articulate the plan and their progress. Customers and investors want to see this progress, and utilities need to develop the right model for their particular business.
>
> John Streur, CEO, Calvert Research & Management.[7]

As discussed, electric power is dominated by electric utilities with largely fossil-heavy generation. That leaves ESG investors with few attractive choices in the energy industry. Firms like Calvert are therefore hungry for power companies that are "climate compatible" – companies with a vision of how to fit into a carbon-neutral economy and a plan to get there. Let us look at four practical examples of companies that took bold steps to reduce greenhouse gas emissions or even started a full transformation to become climate compatible.

Southern Company offers a nice illustration of several tradeoffs discussed so far. Certainly no champion of aggressive action on climate change, Southern Company nevertheless took some interesting steps. Apart from major investments into microgrids and energy storage startups, the company made headlines for two mega-projects: the Vogtle nuclear plant and the Kemper carbon capture and storage (CCS) coal plant. Both have turned out disastrously, with massive cost overruns [26, 27], and the CCS project was eventually abandoned as drastically lower gas prices made it uneconomical [28]. Without the benefit of hindsight though, an ESG investor would have to admit the climate benefits of these investments. Would these bold investments in climate-friendly[8] technologies have made Calvert consider Southern as an investment? The answer is most likely no. Investors look to regulated electric utilities as providers of steady, low-risk returns. Utilities pursuing high-risk endeavors – even for the sake of sustainability initiatives like carbon reductions – are not appealing to mutual funds.

The second example is *NRG Energy*. An independent power producer, NRG Energy is not subject to the regulatory constraints utilities have. Consequently, the company was able to undertake a radical attempt at becoming climate-compatible. On the eve of its transformation, NRG Energy was a giant in US power production

[7] Personal communication. October 28, 2017.

[8] CCS projects typically improve their economic viability by using the sequestered carbon to enhance the output of oil fields, undermining climate benefits. Building large-scale CCS pilot projects helps commercialize the technology, however, with significant potential for long-term climate benefits.

and one of the world's largest corporate carbon emitters. In 2010, then-CEO David Crane decided to get ahead of the inevitable industry changes. NRG was to pivot away from coal-based, centralized generation for wholesale markets and instead expand into renewables, distributed energy, and a shift toward retail markets. Furthermore, in 2014, NRG announced a goal to reduce its carbon emissions by 50% by 2030 and 90% by 2050 [29] – the first US power company to set science-based targets. The transformation soon took shape. NRG became the nation's fourth-largest residential solar developer in 2015, invested around half a billion dollars into a CCS project, and expanded into electric vehicle infrastructure and even portable solar charging electronics. To many observers, the power industry finally had its climate champion.

How did investors react to this transformation? NRG Energy's stock price increased steadily for 2 years between the summer of 2012 and the summer of 2014, climbing from $15 in May 2012 to a peak of $37 in June 2014. Then, many conventional investors got nervous about the direction of the company, and the stock price began a steep descent for the next 18 months. When it fell below $12 in December 2015, NRG's board fired CEO David Crane [30]. For a year, the company tried a similar but more cautious strategy, but investors remained skeptical, and the stock price stagnated. This opened an opportunity for activist hedge funds. In early 2017, two hedge funds acquired a joint 9.4% stake in the company, secured two seats on the board of directors, and pushed NRG to reverse course. The company is now focused on maximizing near-term cash flow and returning to its old business model – a model the current CEO had called "obsolete and unable to create value over the long term" just a few months before [31]. Conventional investors reacted enthusiastically, and NRG's stock price rebounded. In contrast, most ESG investors, wary of NRG even during David Crane's tenure due to the company's still large coal fleet, will likely see their skepticism confirmed and continue to stay away.

Where do carbon-conscious energy investors go instead? One approach is to simply focus on companies that have no power plants, such as transmission companies like ITC Holdings, or distribution-only utilities. With such companies, ESG funds can stay away from controversies and tout their low carbon footprint. Whether such a niche-focused approach supports the industry's transition to a low-carbon future is a different question. A second approach is to look for those companies that have made the most progress in shifting to low-carbon generation.

One of the world's leaders in that regard is the third example, *Energias de Portugal* (EDP). Originally the state-owned utility of a country with just ten million inhabitants, the company began an ambitious international expansion around 1997. As of 2016, EDP earned only about half of its income in Portugal, where the company looks like a standard utility, relying mostly on hydro and coal generation. The other half of EDP's income is derived predominantly in Brazil, Spain, and the United States, where EDP operates as an independent power producer. Globally, in 2016 EDP produced about the same amount of electricity as DTE Energy and Pacific Gas and Electric combined – 70,000 megawatt hours, with a renewable energy share of a staggering 64% [32]. Unfortunately, leadership in renewable power does not necessarily translate to financial performance; EDP's stock has delivered mediocre total returns over the last 5 years (see Fig. 4.7).

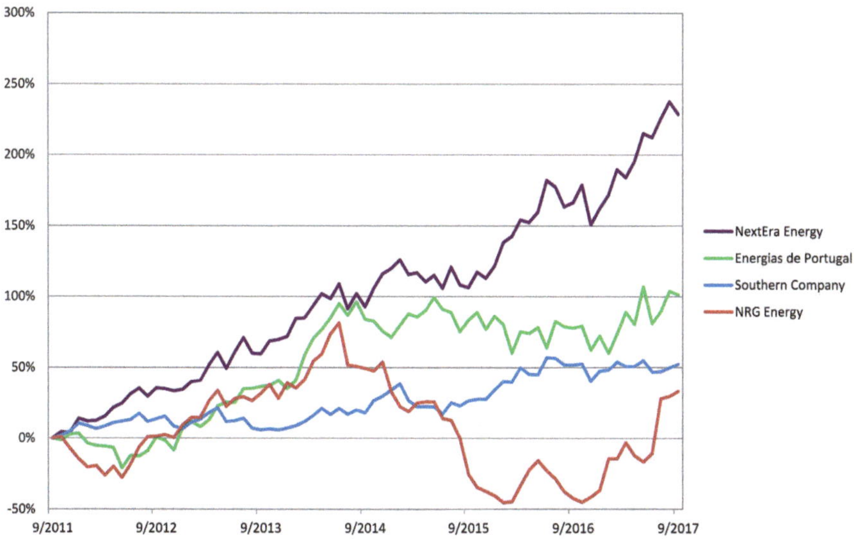

Fig. 4.7 Total return of NEE, EDP, SO, and NRG, 2011–2017. (Based on Calvert data sourced from SNL)

In contrast, *NextEra Energy* (NEE) delivered strong financial results over the same time period, along with an attractive carbon profile. With 20% of generation from renewables, 27% nuclear, and the remainder largely gas,[9] sustainability analysts sifting through vendor-supplied environmental metrics would immediately find NEE. Depending on the fund's sustainability views and vendors' assessment of NEE's social and governance performance, analysts might see it as an investment case without tradeoffs – that is, until he or she has to respond to client letters protesting the fund's nuclear exposure.

As these four examples show, combining financial performance with meaningful sustainability impact is hard and often means accepting tradeoffs between the two. Regarding the discussion at Calvert about the two power companies mentioned in the introduction, it resulted in the decision not to invest in either of them. Their risk of stranded assets seemed too high compared to their pace of transitioning to a low-carbon business. However, we did decide to stay invested in a third power company that has a relatively high carbon footprint. We did not like the fact that it still had a sizable fleet of coal plants, but while that might have deterred other ESG investors, we did see sufficient promise in the company's handling of material sustainability issues. The company's management was interested in discussing its sustainability approach, had publicly committed to ambitious goal to drastically reduce greenhouse gas emissions, and was able to present credible plans and concrete steps how it would achieve those reductions.

[9] Based on Calvert data sourced from SNL.

4.3 Conclusion

Sustainability is a topic that is becoming more important in the wealth management industry. A growing number of mutual funds and other investors are discovering that sustainability does not have to come at the expense of financial performance and can in fact enhance it. There is promise in the growing expectation that investments should benefit both clients and society and that companies pursue profits that are compatible with societal interests.

At the same time, tradeoffs cannot always be avoided, and there are many instances where the interests of investors, clients, companies, and various parts of society at large clash. This is especially true in electric power. In this industry, the list of suitable investments is currently still rather short for many ESG investors. That is unfortunate, because for some of today's key sustainability challenges, the solution hinges on a successful transformation of the electric power industry – led by its companies but supported by investors.

However, where there is risk, opportunity is not far. We observe a gradual increase in the number of utilities taking meaningful steps to reposition their business and address material sustainability challenges. At the same time, the adoption of sustainability considerations is gradually spreading from ESG firms to mainstream investors, and we may well see a growing number of investors combining pragmatism with minimum sustainability criteria. Going forward, electric power companies transforming themselves in a way that balances short- and long-term financial, societal, and environmental needs will find a growing number of keen investors.

Author Profile

Markus Walther Walther covered energy and utilities as an ESG analyst for the ESG investment firm Calvert Research and Management during the writing of this book, before joining Exelon's regulatory finance group in late 2017. At Calvert, he assessed investment risks and opportunities from a sustainability perspective and developed models to compare potential investments. Walther also led shareholder advocacy initiatives engaging companies on environmental and other sustainability issues, especially related to climate change. He is the co-author of the report *Power Forward 3.0: How the largest U.S. companies are capturing business value while addressing climate change*. Walther's previous experience includes stints in private equity, project finance for DTE Energy, and green technology export promotion at the Embassy of Switzerland. Born and raised in Switzerland, he holds an MA degree in International Relations from the Graduate Institute of International and Development Studies in Geneva and dual MBA/MS degrees in Natural Resources and Environment from the University of Michigan. He lives in Washington, D.C. His views do not necessarily reflect Calvert's and should not be construed as investment advice.

References

1. Calvert International Equity Fund, Q1 2018, Calvert, https://www.calvert.com/includes/loadDocument.php?fn=23946.pdf&dt=fundpdfs%27. Accessed 25 May 2018.
2. Caldwell, Tiffany. "Vivint Solar accused of dishonest sales practices by New Mexico attorney general". *Salt Lake Tribune*, March 11, 2018. https://www.sltrib.com/news/2018/03/11/vivint-solar-accused-of-dishonest-sales-practices-by-new-mexico-attorney-general Accessed March 25, 2018.
3. Klas, Mary Ellen. "Insider reveals deceptive strategy behind Florida's solar amendment". *Miami Herald*, October 18, 2016. http://www.miamiherald.com/news/politics-government/election/article109017387.html Accessed March 25, 2018.
4. Colby, Laura. Bloomberg, "Global sustainable investments grow 25% to $23 trillion". July 24, 2017. https://www.bloomberg.com/professional/blog/global-sustainable-investments-grow-25-23-trillion Accessed October 14, 2017.
5. Bresiger, Gregory. Financial Advisor Magazine, "Women, Millennials Want ESG Investments, Advisors Say". October 20, 2017. https://www.fa-mag.com/news/women--millennials-want-esg-investments--advisors-say-35304.html Accessed December 4, 2017
6. Friede, Gunnar, Timo Busch, and Alexander Bassen. "ESG and financial performance: aggregated evidence from more than 2000 empirical studies." Journal of Sustainable Finance & Investment 5.4 (2015): 210–233.
7. Intergovernmental Panel on Climate Change. "Fifth Assessment Report. 2014." Working Group 3, Chapter 1, page 123. https://www.ipcc.ch/pdf/assessment-report/ar5/wg3/ipcc_wg3_ar5_chapter1.pdf Accessed July 9, 2017.
8. U.S. Energy Information Administration, "Capital Cost Estimates for Utility Scale Electricity Generating Plants." Analysis & Projections. November 22, 2016. https://www.eia.gov/analysis/studies/powerplants/capitalcost/pdf/capcost_assumption.pdf Accessed October 10, 2017.
9. National Renewable Energy Laboratory, "U.S. Solar Photovoltaic System Cost Benchmark: Q1 2017.", Figure ES-1. Press. September 12, 2017. Found at https://www.nrel.gov/docs/fy17osti/68925.pdf Accessed on November 22, 2016.
10. Solar Energy Industries Association, Solar Market Insight Reports 2011–2017. https://www.seia.org/us-solar-market-insight Accessed April 2, 2018.
11. U.S. Energy Information Administration, "Estimated small-scale solar photovoltaic generation by state, 2016." Today in Energy. June 1, 2017. https://www.eia.gov/todayinenergy/detail.php?id=31452 Accessed October 10, 2017.
12. Rocky Mountain Institute, "The Economics of Grid Defection, 2017". https://rmi.org/insights/reports/economics-grid-defection Accessed October 24, 2017.
13. Solar Energy Industries Association, "Solar Market Insight" reports for 2011 through 2017. https://www.seia.org/us-solar-market-insight Accessed November 12, 2017.
14. Calvert Research & Management, Ceres, CDP, and WWF. "Power Forward 3.0: How the largest US companies are capturing business value while addressing climate change". April 25, 2017 https://www.worldwildlife.org/publications/power-forward-3-0-how-the-largest-us-companies-are-capturing-business-value-while-addressing-climate-change Accessed January 14, 2018.
15. Rocky Mountain Institute, "Corporate Renewables Deals 2012–2017." Business Renewables Center. http://businessrenewables.org/corporate-transactions Accessed November 15, 2017.
16. U.S. Energy Information Administration, "Utility-Scale Capacity Additions, Retirements and Changes by Energy Source, 2016." Electricity, Table 4.6 https://www.eia.gov/electricity/annual/html/epa_04_06.html Accessed November 15, 2017.
17. Maloney, Peter. Utility Dive. "EIA: Low gas prices set to drive decline in coal generation", January 11, 2018. https://www.utilitydive.com/news/eia-low-gas-prices-set-to-drive-decline-in-coal-generation/514556 Accessed November 15, 2017.

18. Based on analysis of data sourced from S-Network and S&P Dow Jones Indices. Source data available at http://ardour.snetglobalindexes.com/sites/snardour/index-data, and https://www.spindices.com/indices/equity/sp-500 Accessed December 3, 2017
19. Sheppard, David. Financial Times, "Norway wealth fund proposes end to oil and gas investment". November 16, 2017. https://www.ft.com/content/611c2e9e-cad9-11e7-aa33-c63fdc9b8c6c Accessed March 20, 2018.
20. AXA press release, "AXA accelerates its commitment to fight climate change." December 12, 2017 https://group.axa.com/en/newsroom/press-releases/axa-accelerates-its-commitment-to-fight-climate-change Accessed October 8, 2017.
21. JP Morgan Chase & Co. Sustainability. "Environmental and Social Policy Framework." Page 9. Corporate Responsibility. https://www.jpmorganchase.com/corporate/Corporate-Responsibility/document/jpmc-environmental-and-social-policy-framework.pdf Accessed October 8, 2017
22. ING. Newsroom. December 12, 2017. https://www.ing.com/Newsroom/All-news/ING-further-sharpens-coal-policy-to-support-transition-to-low-carbon-economy.htm Accessed October 8, 2017.
23. World Bank, "World Bank Group Announcements at One Planet Summit" News. December 12, 2017 http://www.worldbank.org/en/news/press-release/2017/12/12/world-bank-group-announcements-at-one-planet-summit Accessed October 8, 2017.
24. Fearnside, Philip M. "Environmental and social impacts of hydroelectric dams in Brazilian Amazonia: Implications for the aluminum industry." *World Development* 77 (2016): 48–65.
25. Moran, Thomas. "The environmental and socio-economic impacts of hydroelectric dams in Turkish Kurdistan." (2004).
26. Bade, Gavin. Utility Dive, "Vogtle nuke cost could top $25B as decision time looms". August 3, 2017. https://www.utilitydive.com/news/vogtle-nuke-cost-could-top-25b-as-decision-time-looms/448555 Accessed September 25, 2017.
27. Samuelsohn, Darren. Politico, „Billions over budget. Two years after deadline. What's gone wrong for the 'clean coal' project that's supposed to save an industry?" May 26, 2015. https://www.politico.com/agenda/story/2015/05/billion-dollar-kemper-clean-coal-energy-project-000015 Accessed September 25, 2017.
28. Walton, Robert. Utility Dive, "Southern Co: Kemper coal economics take a hit from low gas prices". February 24, 2017. https://www.utilitydive.com/news/southern-co-kemper-coal-economics-take-a-hit-from-low-gas-prices/436834 Accessed September 25, 2017.
29. Cardwell, Diane. New York Times, "NRG Seeks to Cut 90% of Its Carbon Emissions". November 21, 2014. Page B3. https://www.nytimes.com/2014/11/21/business/energy-environment/nrg-sets-goals-to-cut-carbon-emissions.html Accessed September 27, 2017.
30. Smith, Rebecca. Wall Street Journal, "NRG Energy CEO David Crane Resigns". December 3, 2015. https://www.wsj.com/articles/nrg-energy-ceo-david-crane-resigns-1449151435
31. NRG Energy, Q4 Earnings Call. February 28, 2017. Found at https://seekingalpha.com/article/4050498-nrg-energy-nrg-q4-2016-results-earnings-call-transcript?page=2 Accessed September 27, 2017.
32. Energias de Portugal 2016 Annual Report. Available at https://www.edp.com/sites/default/files/portal.com/documents/Annual%20Report%202016.pdf. Accessed on August 20, 2017.
33. World Nuclear Association, "Nuclear Power in Germany," at http://www.world-nuclear.org/information-library/country-profiles/countries-g-n/germany.aspx.

Chapter 5
Exelon Driving Innovation and the Grid of the Future

Bruce Alexander, Thomas Bonner, William Brady, Christopher Budzynski, Mark Derry, Scott Dupcak, Kimberly Long, David O'Dowd, and John Slocum

Abstract The US electricity grid has withstood the test of time, perhaps more than any other utility infrastructure. It is essentially the same system that we have relied upon for more than 100 years: large central generating stations, transmitting energy over wires to the poles on the streets that ultimately deliver electricity to our homes and businesses. In the future, however, we will experience a very different system for producing and delivering energy. Imagine charging your electric vehicle with energy produced by solar photovoltaic panels on your home, storing excess energy for when you need it, and selling some of it back into the energy system. Through the integration of distributed electricity generation, internet connectivity, and emerging technologies, this future may be just around the corner, provided the regulatory framework that has supported our electricity grid keeps pace with changing customer expectations and emerging technology.

As a Fortune 100 energy company with the largest number of utility customers and among the largest power generation fleets in the nation, Exelon's breadth and scale gives us unique insight into the challenges facing the industry as it adapts to

B. Alexander
Exelon Corporation, Philadelphia, PA, USA

T. Bonner
PECO Energy Company, Philadelphia, PA, USA

W. Brady · K. Long (✉) · J. Slocum
Exelon Corporation, Washington, DC, USA
e-mail: Kimberly.Long@exeloncorp.com

C. Budzynski
Exelon Utilities, Baltimore, MD, USA

M. Derry
Exelon Utilities, Washington, DC, USA

S. Dupcak
Constellation Technology Ventures, Baltimore, MD, USA

D. O'Dowd
ComEd, Chicago, IL, USA

J. Fox, M. Scott (eds.), *Sustainable Electricity II*,
https://doi.org/10.1007/978-3-319-95696-1_5

changing technology and evolving customer demands. This is especially relevant as the pace of technology innovation and digitization of our national energy system continues to accelerate.

5.1 Introduction

For much of the past century, the technology powering our nation's electric grid has not dramatically changed. Large, centralized power stations produce electricity, and utilities deliver it through the transmission and distribution (T&D) system wires to homes and businesses. As our communities expanded, we simply built more power plants and installed more wires. Customers' primary interaction with their utility came at the end of each month when they paid their bill. Today, technology and customer demands are changing at an unprecedented pace. A decade ago, no one could have imagined how companies like Uber, Amazon, Netflix, and Airbnb would tap into the digital revolution, creating virtual marketplaces and fundamentally altering the way consumers use common services that had long been the domain of traditional brick and mortar businesses. Similarly, the rapid adoption of smart grid technology, combined with increasingly sophisticated energy efficiency systems, new sources of distributed energy, and the profound changes brought about by abundance of natural gas are strong indicators that the electric utility industry is poised for change as a result of the technological forces that are reshaping our world.

Emerging technology is changing the way utilities produce, store, and manage energy use, posing new challenges to regulators and grid operators as they race to keep pace with the technology integration. It is also changing the way consumers interact with the energy grid as utilities provide new products and services to meet demand for clean and reliable energy.

The trends affecting this transformation (Fig. 5.1) include:

- *Evolving Consumer Behavior and Expectations* – Customers want greater control over their energy use and choices. They are increasingly considering energy generation and use as an integral part of their lives and businesses by embracing local generation, such as residential and commercial solar, by implementing energy efficiency measures and by utilizing technology such as smart phones and tablets to track and manage home and business energy usage and monitor it remotely.
- *Abundant and Affordable Natural Gas Supply* – Over the past 10 years, shale gas drilling technologies and affordability of natural gas prices have encouraged greater use of this fuel for power generation. The dominance of natural gas availability has contributed to more affordable electricity prices and changes in production margins, making it economically more challenging for some generating plants to compete, including putting some emission-free sources like nuclear at a competitive disadvantage.
- *Flat to Low Demand Growth* – In addition to low natural gas prices, power prices have also been impacted by sluggish growth in the demand for electricity. Over

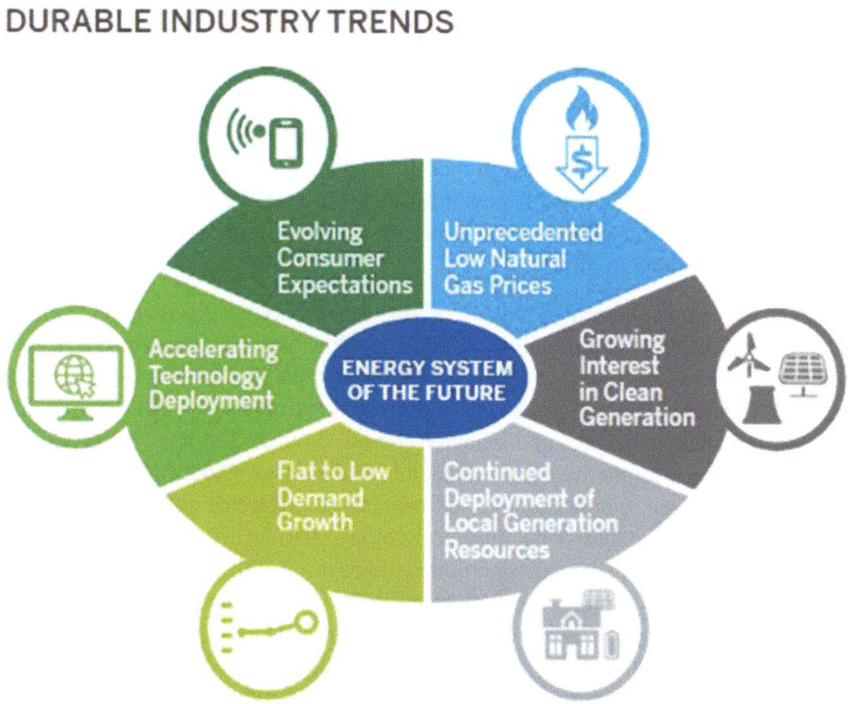

Fig. 5.1 Electric power industry trends

the past decade, power suppliers are seeing flat to very low growth in demand as homes and businesses adopt energy efficiency measures that lower their energy use. The lingering effects of the 2008 recession also have played a role, and these developments represent a fundamental shift in market dynamics compared to the higher demand growth rates of prior decades.

- *Continued Deployment of Local Generation Resources* – The deployment of local generation such as solar, wind, or battery storage in residential and commercial applications continues to increase. This is largely in response to customer interest in renewable energy as well as options to increase local grid reliability through technologies such as battery storage. Renewable energy deployed by local utilities, competitive generation companies, and private customers is supplanting some conventional generation.
- *Accelerating Technology Deployment* – The centralized generation and T&D system, though fundamentally needed to supply and distribute electric power, is also transforming. An intelligent electric network, enabled by two-way communication technologies and the expanding "Internet of Things" ("IoT"), is emerging to create a smarter and more efficient power grid. Both regulated utilities and third parties are deploying new technologies such as advanced metering infrastructure (AMI) that provide options for more efficiently monitoring and managing energy usage.

- *Growing Interest in Clean Generation* – There is sustained and growing public interest in reducing the impacts of energy usage on the environment, whether related to climate change and carbon dioxide emissions or other air pollutant emissions such as nitrous oxide and sulfur dioxide that can contribute to unhealthy air quality. In the absence of federal energy policies to address emissions, many states are moving on their own to implement policies that promote affordable, clean, and reliable power for their citizens. While federal policy initiatives have fluctuated over time, customers' long-term interest in clean generation has continued to grow.

Electricity now represents a suite of products and services; it is no longer simply a commodity product and functions less as a pipeline that delivers electrons from a central generating station to residential and commercial customers and more like a network of interconnected producers and consumers of energy. Nowhere is this more evident than in the area of distributed energy resources (DER), which continue to expand along with the fundamental decline in the cost for rooftop solar and battery storage, as well as new and diverse financing products that make adoption easier and more affordable. Changing technology also provides consumers and businesses with unprecedented opportunities to customize their energy sources and improve their ability to make better, more informed choices about their energy management. Aggregated demand response programs such as Google Nest's "Rush Hour Rewards" [1] compensate and encourage customers to reduce energy use during times of peak demand, such as on a hot summer afternoon. Similarly, many utilities offer programs that compensate customers for reduced energy use during peak demand events, and initiatives like Google Sunroof [2] give customers the ability to easily determine the potential for rooftop solar for a building.

Wind and solar resources are also being developed in many regions of the country more quickly and on a larger scale than anticipated. This creates challenges for grid stability in areas where the T&D infrastructure cannot handle the additional load and lacks sufficient energy storage capacity. For example, the affordability and access to solar assets in Hawaii created a situation of grid instability as the T&D infrastructure could not accommodate the significant increase in individual customer rooftop solar installations. The increase in installations was due, in part, to the desire of customers to manage their electricity usage and cost [3] coupled with the state's goal to supply 100% of its energy from renewable energy sources by 2045 [4]. Similarly, a recent California state senate proposal, SB100 [5], has set a goal to meet all the state's electricity needs with renewable energy sources by 2045. This would require significant upgrades to the existing T&D system along with the need for additional energy storage capacity.

Like other companies in the electric utility sector, Exelon is adapting its business to keep pace with changing technology and customer expectations for affordable and clean energy alternatives (Fig. 5.2).

From 2016 through 2020, Exelon plans to invest $25 billion into our utilities, with a focus on technology deployment aimed at incorporating energy products and services that provide greater benefits to customers. These investments will also protect critical infrastructure and facilitate the implementation of smart grid capability, com-

Fig. 5.2 "As the only Fortune 100 company in the electricity sector, we have a unique opportunity to lead the energy industry in the exploration, development and deployment of the next generation of clean, diverse energy technologies." Chris Crane, President and CEO of Exelon [6].

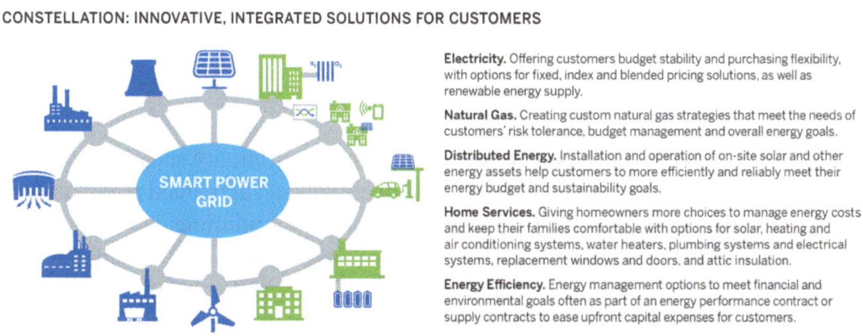

Fig. 5.3 Constellation services

munication networks, and data management systems enabling two-way communication between electric utilities and customers. We are reimagining how we produce and deliver energy to customers through the advancement of microgrids, and we are developing new tools and policies to assist our customers in gaining access to renewable energy, including DER such as rooftop solar photovoltaic (PV) installations. Constellation, Exelon's competitive energy products and services business, has worked with business and public sector customers across the country to deploy more than 400 MW of DER at customer locations, such as solar, cogeneration, standby generation, fuel cells, and battery storage. Figure 5.3 highlights the diverse services, including DER, offered by Exelon's Constellation business.

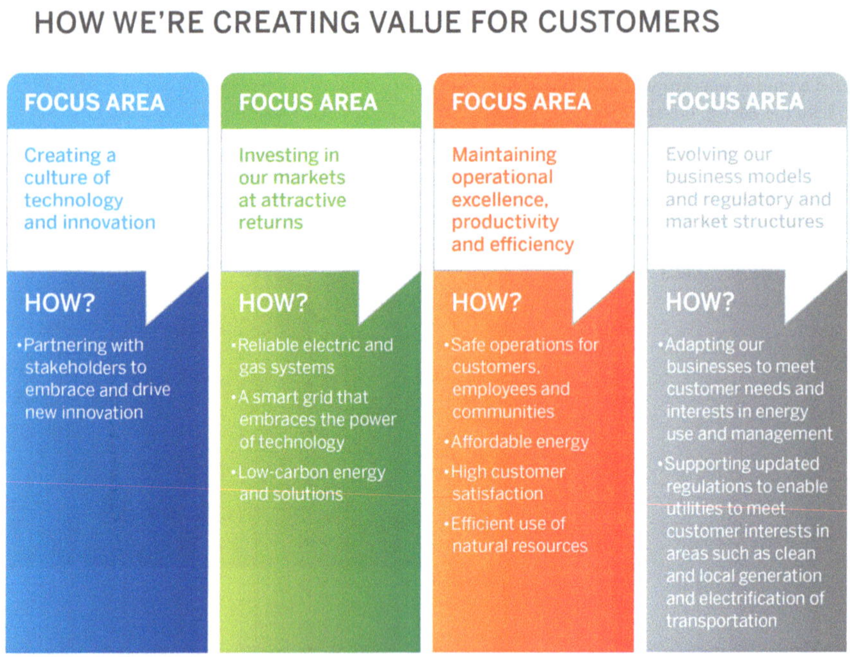

Fig. 5.4 Exelon Strategic Plan Focus Areas

To help inform the development of forward-thinking electricity and energy public policy solutions, Exelon continues to work with elected and appointed policymakers and stakeholders at all levels of government. Our policy priorities are focused on promoting the advancement of a safe, reliable, and clean energy future through greater customer access to renewables and DER. They seek to ensure that all technologies are fairly valued in competitive markets, while also preserving the security and efficiency of the nation's electric grid and encouraging investment and innovation as reflected in key focus areas of our strategic plan, detailed in Fig. 5.4. The deployment of smart assets and equipment like AMI to build a more intelligent electric grid network offers better visibility into the performance and status of the grid. It also provides the ability to utilize data to better understand customer needs and more efficiently manage the grid. The electric grid of the future will not resemble the one we have now; it will be more interactive and accommodating of emerging technologies. It will require greater coordination among electric utilities, regulators, technology companies, and customers and has inherent challenges, many of which relate to the current regulatory structure, equitable pricing, equitable investment cost recovery, and competition. It represents a massive transition for the system, and the transformation is still in its infancy.

5.2 Engaging in Grid Modernization Regulatory Dialogue

To address the combined effects of new technologies, new business models, and evolving customer interests, utilities and state public service commissions are reassessing historic regulatory frameworks that govern the operations and funding of regulated utility T&D systems. In some instances, state and federal jurisdictions charged with implementing public grid modernization policies are not as well aligned with changing public policy priorities, such as in the case of battery storage. Traditionally, regulated T&D utilities have not been permitted to own generation sources; if regulations characterize battery storage as a generation source when it is releasing energy, it can limit the extent to which utilities participate in investing in this technology even when it benefits grid reliability and improves customer service. Further, policy changes in one area of regulations can create unintended consequences or disconnects in other aspects of the business that are regulated for unrelated reasons. For example, utilities must manage their operations and investments to address multiple energy, market, and environmental regulations. If mandates in one area, such as environmental regulation or renewable portfolio standards (RPS) to reduce carbon emissions, do not integrate and align with regulations in another area, such as market regulation mechanisms that value low-carbon energy and support incorporation of DER, consequential regulatory disconnects can occur, resulting in inefficient and conflicting outcomes. In states where aggressive RPS goals exist, such as Hawaii and California, the influx of solar and wind generation has created new and more challenging operational conditions as more intermittent energy supply is created. These conditions are occurring, in part, as a result of operating the system in a way in which the system was not designed due to limited energy storage resources. In the case of California, solar and wind power productions have been curtailed, and the state has paid Arizona to accept excess electricity to avoid overloading of the grid because more electricity has been available than its residents have been able to utilize [7].

5.2.1 Regulatory Challenges and Opportunities

One of the challenges facing the industry is determining how to incorporate variable energy sources – such as wind, solar, and battery technology – into a regulatory system and infrastructure that was designed for conventional energy sources that can be easily ramped up and down to match customer demand. A challenge of renewable and DER technologies is that their variable nature can be disruptive to the traditional grid design and function under the current regulatory and mechanical framework. This presents an opportunity for utilities and public service commissions to collaborate on developing new regulatory and infrastructure solutions that will facilitate the introduction of more DER resources.

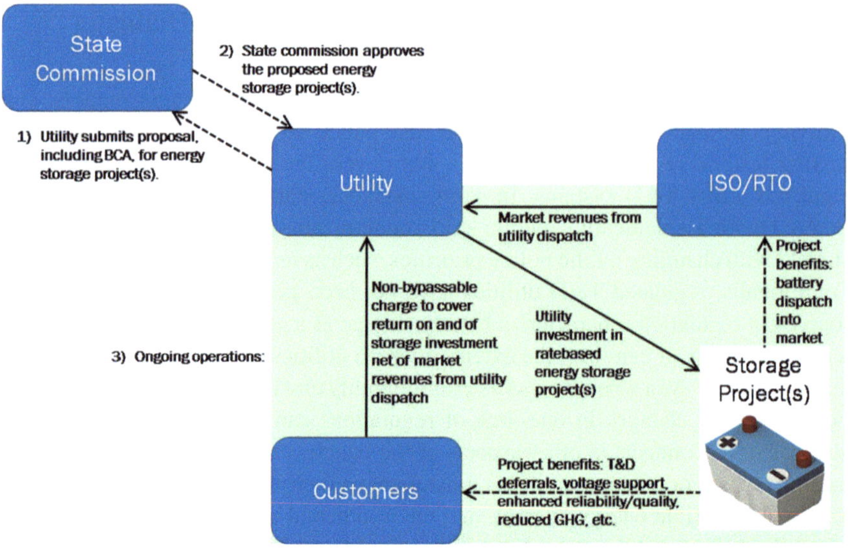

Fig. 5.5 Conceptual regulatory framework for energy storage grid integration

Over decades, states in which Exelon has utility operations have established varying regulatory structures and priorities. Restructured states, regulated by state commissions, separate electric distribution from generation supply, which operates in a competitive market and is regulated by the Federal Energy Regulatory Commission (FERC). The challenge remains that many of the emerging technologies have functions on both sides of this division, including energy storage, distributed generation, and microgrid technologies, all of which provide a mixture of power supply as well as demand, and this blurs the lines between traditional utility wire company services and competitive customer and energy services businesses.

To address this, Exelon is working with states to identify and advance legislative and regulatory changes to modernize the regulatory framework and better serve customers. Opportunities include enabling investment in areas where state regulation may currently limit utility investment and ownership, such as in the deployment of urban microgrids, energy storage applications, and the provision of solar options for customers. As the regulatory framework continues to evolve, we continue to deploy DER consistent with existing requirements and restrictions and await potential changes whereby the traditional T&D structure may transition to a service provider model, an energy entity that provides services to a retail or end-use customer [8], which incorporates DER resources. Figure 5.5 illustrates a conceptual regulatory framework of how energy storage projects could be incorporated into the grid.

Key to incorporating emerging technologies and maintaining reliability of the grid is ensuring that the value of the grid that is used by all is recognized and adequately funded. As customer adoption of rooftop solar and other DER technologies grows and the customer base becomes further differentiated, rate structures, which establish revenue returns to provide for maintenance of the grid that are based on how much energy a customer uses (e.g., dollar per kilowatt hour), must be updated to a rate design where customers pay their fair share for the use of the grid. This would ensure that customers who get most of their energy from a rooftop solar installation, for example, still contribute to the cost of maintaining the grid. California and Hawaii have been grappling with this sort of electric rate reform for some time. California is attempting to adjust electric rates such that each user pays for the actual cost of service, establishing a structure of rates based upon when customers use energy and how much energy is used at once. Since 2001, due to California's four-tiered rate structure, most of the burden of electricity rate increase was born by only a small portion of the customer base [9]. In Hawaii, many customers who can afford to install rooftop solar and battery storage to combat rising electricity rates have done so, shifting the burden of electricity rate increases to those who remain connected to the grid [10].

A recent example of industry successfully working with policymakers to implement new laws and regulations to address energy challenges is the passage of the Illinois Future Energy Jobs Act (FEJA) [11]. The FEJA addressed two critical needs: First, it preserved two of the state's most efficient nuclear plants, which faced imminent closure as a result of distorted market signals and energy policies that did not value nuclear power for its many attributes, notably that it generates electricity around-the-clock with no air pollution or other greenhouse gas (GHG) emissions. Second, it addressed flaws in the state's RPS, which held back investment in new sources of wind and solar energy. Exelon worked with a broad coalition of more than 200 business, labor, environmental, and faith-based organizations and other groups to promote the FEJA. The legislation:

- Enhances the state's RPS and provides for billions of dollars in investment in clean energy and energy efficiency across the state.
- Creates a zero-emissions credit (ZEC) program that levels the playing field for all clean sources of electricity, including solar and wind, by placing a value on zero-carbon energy. This allows the electricity from Illinois' nuclear facilities, which account for more than 90% of the state's zero-carbon energy, to contribute to Illinois's energy security.

Prior to the establishment of the FEJA, two of Exelon's nuclear facilities in Illinois faced imminent closure, the state's RPS needed a legislative fix, and there was a significant growth opportunity for energy efficiency programs. Absent proactive bipartisan action, the state was at risk of deteriorating its strong economic position and losing its advantage as a clean energy leader. Illinois energy policy stood at a crossroad. Throughout the 3-year policy development, Exelon remained a key party to policy discussions and negotiations. On numerous occasions, Exelon representatives provided expert testimony to Illinois House and Senate committees on several topics including energy efficiency, the value of distributed energy resources to the electric grid, the environmental and economic benefits of nuclear power stations, and much more. The result, FEJA, was the product of collaboration among environmental, consumer, labor, and energy interests evaluating tradeoffs among their respective objectives and identifying common goals. The FEJA authorized increased deployment of energy efficiency measures, updated the state RPS to provide for more in-state renewable energy, and preserved two of the state's most efficient nuclear plants that were facing imminent closure by creating value in the energy market for zero-carbon energy. Among many important provisions, the FEJA also provides for decoupling of utility revenue from energy usage, ensuring adequate funding for the T&D system even as energy efficiency may reduce levels of delivered power. It also provides for rate-based solar rebate programs for commercial and community solar projects.

Following creation of the FEJA, the ICC adopted a resolution launching Illinois' Utility of the Future proceeding, dubbed the "NextGrid" Grid Modernization Study [12]. ICC Chairman Brien Sheahan announced the 18-month process and stressed that it would be consumer-focused and leverage the benefits of the restructured energy market. According to the ICC, the initiative will identify and explore future technological advancements and utility and regulatory models. The process is intended to inform Illinois policymakers on the issues and challenges associated with the quickly evolving energy landscape and provide policy recommendations to the ICC and Illinois General Assembly as they are identified.

Efforts similar to that in Illinois are also underway in Maryland and the District of Columbia (DC) with the PC 44 [13] and the Modernizing the Energy Delivery System for Increased Sustainability (MEDSIS) [14] initiatives, respectively. Maryland's PC 44 grid modernization seeks to address topics of rate design, DER implementation, and integration to the grid, energy storage, interconnection processes, and distribution system planning. The MEDSIS initiative in DC proposes to identify technologies and policies that will modernize the electric grid while maintaining the reliability of the system, providing value to customers, increasing the efficiency of the system, and ensuring the sustainability of the system for years to come. The Exelon utilities in each jurisdiction support the perspective that energy storage resources should not be treated as electric generating facilities to help avoid uncertainty in regulatory treatment. Exelon also supports the ability of utility-owned energy storage resources participating in energy markets in a way that maximizes the value of the system, while reducing total system cost for customers such as by providing ancillary services (i.e., scheduling and dispatch, voltage control, and system protection) that support the transmission of electric power from seller to purchaser while maintaining reliable operations of the system.

5.3 Driving Innovation Within

Beyond our efforts to enable customers to adopt emerging energy supply and management technologies, we are also seeking to incorporate innovative approaches for improving our business into how we operate internally. One way we are doing this is through collaborative relationships with other private companies, academia, government institutions, and national labs. These partnerships facilitate collaboration on very early-stage research, not only between Exelon and our partners in each category but also across categories to develop well-rounded technologies that tackle long-term challenges. We share our vision of the future with our partners, who add their expertise and input. Technologies that may play a role in the future are selected for investment through Exelon's venture investing group. The group seeks opportunities to balance the uncertainties of regulatory reform and its impacts on the incorporation of DER by electric utilities into the existing T&D infrastructure.

Constellation Technology Ventures (CTV) invests in growth-stage companies representing innovative technologies and business models that complement, or may disrupt, Exelon's core businesses with the goal of providing innovative solutions to Exelon and our customers. Investments through CTV comprise a wide range of new technologies, including electric vehicles, distributed generation, energy storage, renewable generation, and intelligent buildings. In addition to monetary investment, Exelon provides expertise and resources to further collaboration. Following investment, portfolio companies engage with CTV's New Business Incubator, where a specialized team facilitates commercialization of CTV investments and other modern technologies within Exelon business units. Two CTV portfolio companies with which the New Business Incubator team has engaged include Proterra [15] and Stem [16].

Proterra is a world leader in the design and manufacture of zero-emission vehicles that enable bus fleet operators to reduce operating costs and deliver clean, quiet transportation to the community. The Proterra Catalyst™ electric transit vehicle, manufactured in the United States, was designed to be the most flexible platform for the electrification of transit systems around the world, offering customizable options that provide a wide range of configurations based on individual routes, such as transit and campus implementation. When a Proterra bus replaces a diesel bus, an estimated 244,000 pounds of carbon dioxide (CO_2) are avoided over its operational lifetime. Proterra has now sold more than 380 vehicles to 36 different municipals, universities, and commercial transit agencies in 20 states across the United States. In 2016, King County Metro in Seattle and adjoining communities placed an order for 73 Proterra buses, the largest order of its kind in North America. Exelon pursued the project with Proterra because the development of the electric bus represents a vision of the future for mass transit options. A typical Proterra bus is shown in Fig. 5.6.

5.3.1 Energy Storage

Exelon is investing in energy storage and regulatory reforms that will enable the benefits of energy storage to take on a greater role in electricity supply. Such technology can contribute to a future system that is more dynamic and resilient and where customers have more choice and control over their energy use while still maintaining electric reliability in a cost-effective manner. Energy storage technologies offer benefits such as the potential to reduce costs, improve grid reliability, support clean energy alternatives such as solar, and provide for technology sector

Fig. 5.6 Proterra Catalyst™ electric transit vehicle

job creation. Energy storage also presents a new set of opportunities to modernize the regulatory compact as part of the grid modernization landscape. Exelon is developing a number of customer and grid storage options, such as the Stem application, a behind-the-meter energy storage technology, which will require changes to the regulatory framework to allow utilities to own, operate, and earn an appropriate return on their energy storage investment. Grid modernization regulatory efforts to incorporate energy storage technologies such as batteries could also encourage the development and use of other storage technologies, such as thermal energy storage. This evolution can facilitate a situation where electricity may be used in off-peak hours to produce ice that can later be used to reduce the electrical demand of cooling systems during peak hours. Additionally, energy storage in a substation or on a distribution feeder, which has large electric current capacity, could be utilized to enhance reliability of the distribution system. Energy storage can also defer the need to increase generation capacity in that portion of the distribution system while continuing to meet customer load demand. The ability to utilize DER and energy storage technologies can improve system reliability at a lower cost to the customer than traditional technology.

Energy storage can also support managing and mitigating the impacts to other customers from intermittent changes in voltage on the system since solar or other DER energy production varies due to sun and wind conditions or other factors. By discharging during peak hours, battery storage can reduce the need to produce additional electricity from centralized generation sources while ensuring adequate energy availability during times of peak energy demand.

Regardless of the type and amount of energy storage installed on the distribution system, the power grid must be capable of handling the highest peak energy usage throughout the year, even though the average usage throughout the year is roughly one-half to two-thirds of the peak. Because batteries can charge and discharge power within seconds, they can be utilized to respond quickly to short-term increases in electricity demand. In some states, current regulations challenge the ability of utilities to own electric generation sources such as battery assets, since the batteries can be considered generation sources when they are releasing energy to the grid.

However, states like California, Oregon, Massachusetts, Ohio, and Florida have enacted legislation to facilitate energy storage deployment.

As costs of battery technologies are rapidly decreasing, recognition for benefits associated with energy storage such as voltage and reliability stabilization and cost savings to management of the grid and its customers are increasing. A battery project may act as a grid asset, allowing a utility to defer or avoid other distribution system investments such as the upgrade of a transmission line. For example, historically, a persistent overload condition would have been addressed by performing a comprehensive upgrade to a substation. However, utilizing a battery storage alternative tailored to address the forecasted overload could cost significantly less and could potentially defer the need for the substation upgrade indefinitely.

Exelon's BGE utility is planning to utilize batteries in lieu of substation and transmission upgrades for load support. Through the use of batteries, they will be able to more cost-effectively provide voltage support and avoid flicker, a visible change in brightness of a lamp due to rapid fluctuations in power supply voltage. While this example represents only a deferment of the traditional investment, this type of alternative does provide an opportunity to manage customer electricity price increases for a longer period of time. For these reasons, Exelon supports an approach to energy storage that allows utilities to own and utilize batteries and their associated control systems as grid assets coupled with operation of those assets in a manner that maximizes their value for customers and society to not only maintain reliability of the grid but also manage costs for customers.

Stem, a company that serves as a leader in behind-the-meter energy storage applications, delivers cost savings to customers through peak demand reduction programs and to utilities through demand response programs. The Stem system combines industry-leading lithium ion batteries with Stem's proprietary software for analyzing and predicting demand. The technology helps businesses and public institutions automate energy cost savings, respond to changing energy use, and support a sustainable grid. Stem shifts energy use away from times when it is most expensive, learning each building's energy patterns to maximize the value of on-site storage. The benefit of Stem to customers is having the ability to take advantage of energy storage with no upfront costs or disruption in service. Another benefit of Stem is that its customers form a network that helps utilities create a more efficient, sustainable, and resilient electrical grid. Stem's network includes more than 100 MWh across 600 facilities, making it the largest commercial-scale energy storage network in North America. This project was pursued by Exelon to address the dual focus of providing customers with a DER technology to address high demand charges while also providing stability to the grid.

5.3.2 Solar

In the case of solar, state support for the technology is not uniform. Some states, such as California, North Carolina, South Carolina, Massachusetts, and Hawaii, have high allocations for solar in their RPS programs, tax incentives, or favorable net metering treatment. This is not the case in jurisdictions such as Illinois, Pennsylvania, Indiana, Louisiana, and Ohio, which lack tax and other policies that encourage solar installation. In addition, state handling of solar renewable energy credits (SREC) can also influence implementation of solar technology. SRECs are tradable energy commodities that provide verifiable certificates that represent the renewable electricity attributes associated with 1 MWh of electricity generated from a qualified solar energy resource. Because SRECs are sold separately from the electrical energy from which they are co-produced, some states like Pennsylvania limit the purchase of SRECs from out-of-state sources, resulting in pricing that can vary due to supply and demand [17]. In addition, the investment tax credit (ITC) provides an investment incentive for those who invest in new solar generation with a tax credit, a feature that is planned to be phased out by 2023. The interplay between the ITC and SRECs coupled with cost recovery remains an important financial driver for utilities as they consider expansion into solar markets. Exelon has performed evaluations of solar installations by size, including small residential/commercial and pole mount types, medium community/commercial and car port types, and large grid/utility scale types. Exelon supports the deployment of solar technology, in part, because it is one solution for meeting customers' expectations for clean energy. However, based on considerations such as operating and maintenance expenses and cost per watt installed, our business focus is on large-scale solar implementation, often referred to as utility scale, as it is more cost-effective and attractive to our investors at the present time. Figure 5.7 presents some of the advantages and disadvantages to the use of solar power.

5.3.3 Microgrids

A microgrid is a small network of electricity users with a local energy generation source that is connected to a larger centralized grid but can function independent of the overall power grid. Microgrids also represent an emerging technology that offers a variety of benefits, including the ability to provide a reliable source of power for critical community assets such as hospitals or fire stations during major outages and emergencies. However, integrating these systems into the grid poses challenges. Gaining regulatory support for the concept and its implementation has been difficult in many states due to concerns about how costs are allocated to customers. Because microgrids represent electricity distribution systems that contain electricity loads as well as DERs – such as generation sources and storage devices – they can be operated in a coordinated manner with the rest of the electric grid or in an isolated fashion. For this reason, regulatory changes are needed in many states to address the ability of utilities to own and operate these types of systems. An example of a typical microgrid is shown in Fig. 5.8.

Exelon supports the development and implementation of microgrids by utilities because of our experience with the operation of the electric grid and advantages of microgrids to provide stability, reliability, and security to the T&D system. Exelon's utilities are working with equipment suppliers and government to design and

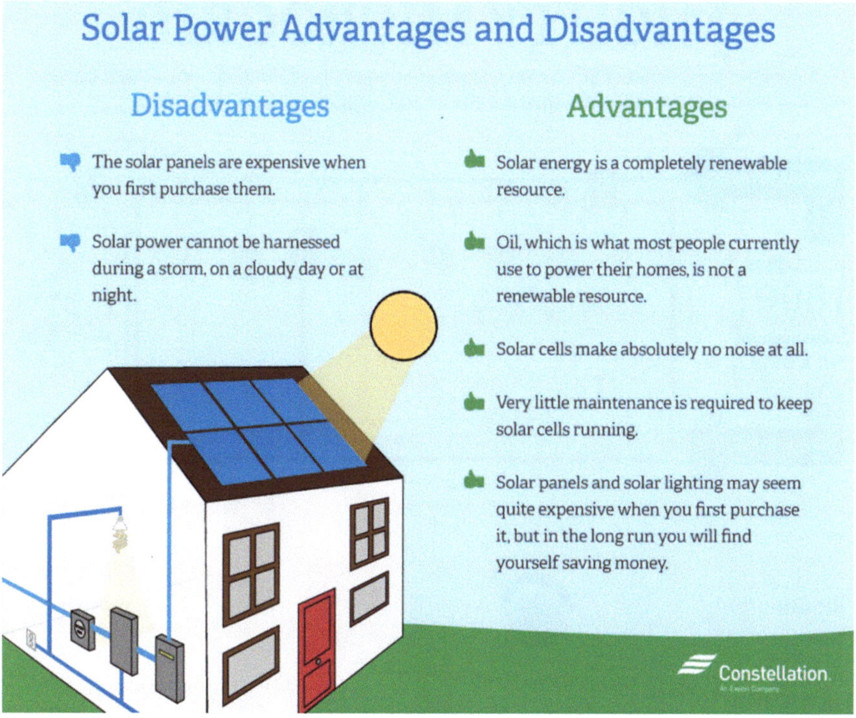

Fig. 5.7 Solar power advantages and disadvantages

develop microgrids for improved resiliency and integration with the grid. To establish a legislative and regulatory framework to promote the development of microgrids and innovative energy storage applications, PECO has worked with Pennsylvania representatives on the development of HB 1412, the Energy Resiliency and Emergency Preparedness and Response Act [18]. Through a bipartisan coalition of legislative committee members, HB 1412 is moving through committee and toward the house floor as a crucial step in supporting grid resiliency and critical emergency preparation and response activities.

Exelon, through Pepco Holdings, Inc., is working on a microgrid project with Chesapeake College [19] in Maryland to handle critical loads during outages and incorporate on-site solar and storage assets. The microgrid project is comprised of a combined 1.76-MW solar array and battery storage on its campus. An initial challenge with this project was having sufficient capacity on the grid to accept the generation from the college's solar array due to another solar array in the area taking up existing capacity on the grid. The addition of advanced inverters that receive utility control signals and battery capacity enabled the system installation as well as the ability of the college to generate power for the grid and the opportunity for the college to take advantage of storage capacity for backup power in emergency situations. In addition to the Chesapeake College project, Exelon's PECO and ComEd utilities are engaging with interested stakeholders to determine how to best proceed with the implementation of microgrid technology across their territories.

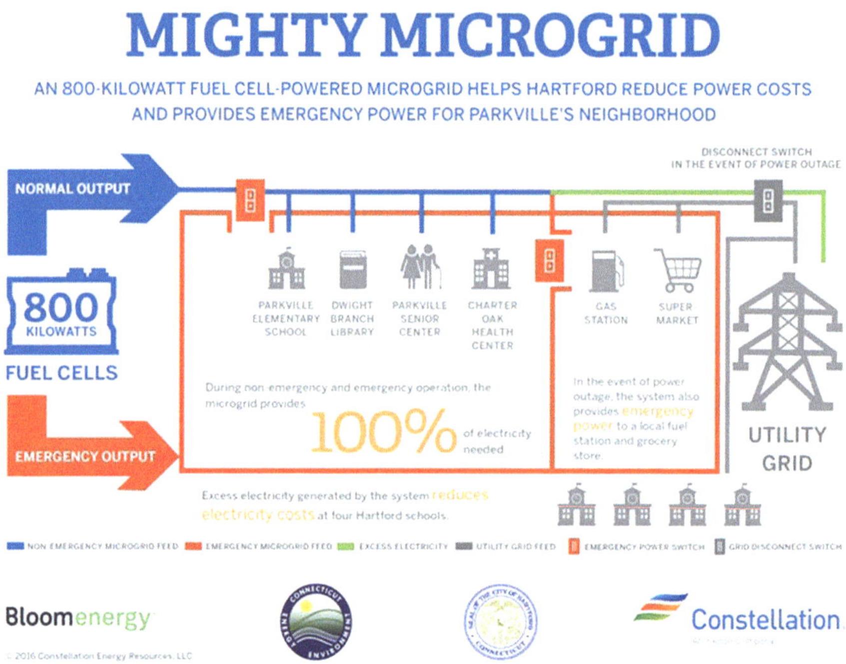

Fig. 5.8 Hartford, CT microgrid

5.3.4 *Electric Vehicles*

Adoption of electric vehicles (EVs) is partially limited by accessibility to public charging stations. Determining how to fund EV charging stations through an electric utility can pose a regulatory challenge, similar to the deployment of solar, energy storage, and battery technologies. Generally speaking, it is hard to convince customers who have access to mass transit and public transportation to subsidize suburban customers who may have better ability to afford an EV. EV implementation is also dependent upon the status of state regulations encouraging a shift to their use over traditional vehicles. For example, California is very EV-friendly, as it is a zero-emission vehicle state and has incorporated goals into state regulations for their use. Outside of California, there are few states that have a specific target for EVs.

Exelon utilities are developing plans to promote EV adoption and integration within the grid through infrastructure investments and incentives. Specific initiatives being considered include installation of public chargers in the BGE and ComEd territories and rebates for home and business chargers in the BGE territory. Other initiatives in development include an EV smart charger pilot project and development of residential rate options in PHI territory. Serving some of the country's largest metropolitan areas, Exelon utilities also are engaging to help deliver the benefits of electrified transportation into the transit sector. In 2016, PECO partnered with the Southeast Pennsylvania Transit Authority (SEPTA) and electric bus manufacturer Proterra to demonstrate the benefits of electric buses during the 2016 Democratic National Convention in Philadelphia. Throughout convention week, more than 1000 elected officials, convention delegates, and local residents toured and took demonstration rides on the Proterra bus. PECO is supporting SEPTA's addition of 25 Proterra buses on its system in late 2017, working with the transit authority to identify solutions that ensure reliable charging infrastructure is available to support the initiative.

In addition, our Pepco utility will be introducing a pilot program to investigate the reliability impacts of charging behaviors and testing methodologies to incentivize customers to charge their EVs when distribution system demands are typically low. One challenge with EV charging implementation in the Pepco DC territory is that a significant portion of DC residents live in multi-dwelling units and only have on-street parking access. It is likely that these residents would prefer direct current fast-charging options as they do not have the ability to charge their vehicles for 8 h or more in a reserved space. The Pepco pilot program will help identify additional challenges and opportunities to support EV charging access in DC and provide data that can be used by other electric utilities.

To assist with the provision of new value-added service to customers, electric usage associated with EVs helps to stabilize overall costs of maintenance for the electric grid, and it provides societal benefits in the form of lower GHG emissions. PECO has worked with Pennsylvania lawmakers on the Clean Transportation Infrastructure Act [18], which would establish a state goal for expansion of transportation electrification and require the development of regional infrastructure plans by electric utilities in the state's major metropolitan areas. The bill has attracted bipartisan support in the Pennsylvania General Assembly, and consideration is expected in the latter part of 2017.

5.4 Conclusion

The future of the electric grid will continue to be influenced and impacted by innovative technology. We are witnessing the acceleration of the pace of technology innovation and digitization of our national electric system which now functions as a suite of products and services that facilitates greater control and management of electricity use by customers. We are shifting from the traditional centralized power generation model to a decentralized one that includes DER such as solar and battery storage, among other technology options. However, the variable nature of some DER technologies can be disruptive to the traditional grid design under the current physical and regulatory framework, and this presents an opportunity for the electric power industry to work collaboratively with public service commissions and other regulatory and policy organizations to develop new regulations and infrastructure solutions to facilitate the incorporation and use of new technology. Some of the keys to successfully integrating emerging technologies and achieving regulatory reforms include:

1. Enabling electric utilities to offer a variety of energy products and services, including distributed generation and storage
2. Utilizing data analytics to gain a deeper understanding of customer expectations
3. Creating innovative regulatory structures that encourage investment and facilitate equitable pricing of new products and services

Piloting and deploying new technologies such as DER, energy storage, and microgrids will provide greater flexibility to respond to the changing customer demands and dynamics of the grid, and collaboration of technology implementation with regulatory officials, stakeholders, and customers will ensure the continued availability of affordable, reliable, and clean electricity and support the continued evolution of the electric grid. Through our company business plan, we are embracing a culture of technology and innovation, investing to improve grid reliability and customer service, demonstrating operational excellence, and advocating for regulatory and market structures that are needed to bring about the next-generation energy grid.

Author Profiles

Bruce Alexander Alexander currently manages several of Exelon's corporate sustainability reporting programs, such as the annual corporate sustainability report, and engagements with investors and other stakeholders regarding sustainability issues and performance, including corporate responses to key surveys such as the Dow Jones Sustainability Index survey. Alexander is also involved in business planning for the Exelon Corporate Strategy and Innovation and Sustainability Group and manages Exelon's environmental management system focus area that reviews emerging regulatory and legislative issues related to environmental and sustainability issues.

Thomas Bonner Bonner has primary responsibility for analyzing the impact of federal, state, and local policy challenges on the company and leading internal teams to develop policy and business solutions. During his time with PECO, Bonner has focused on both policies and programs advancing new technologies, including serving as the lead author of PECO's successful $200 million Smart Grid Investment Grant application with the US Department of Energy. Following the grant award, he organized the company's grant compliance assessment and implementation planning project and functioned as Project Manager for PECO's external partnership programs. Prior to joining PECO in 2007, he served as Manager, Congressional Relations for MidAmerican Energy Holdings Company for more than a decade, and as Associate Director of the Commonwealth of Pennsylvania's Washington, DC office. From 1991 to 1995, he worked in the US House of Representatives as a senior legislative staffer. Bonner holds a Master of Planning degree from the

University of Virginia and a Bachelor of Arts degree from the University of Pennsylvania.

William Brady Brady leads Exelon's safety and environment functions and is responsible for ensuring effective governance and oversight for these areas throughout Exelon, including the company's generation, electric and gas utilities, and Constellation energy services businesses. He supports the senior leadership team with the development and execution of Exelon's sustainable business and environment strategies. His organization is also responsible for the company's public environmental and safety reporting, including the annual Exelon Corporate Sustainability Report and the Dow Jones Sustainability Index and Carbon Disclosure Project questionnaire responses. During his 35-year career in the energy sector, he has worked on a wide range of issues and energy technologies, including nuclear power, cogeneration, biomass, renewable energy, and distributed generation. He began his career in engineering, has experience with environmental marketing and developing business strategy, and more recently has been focused on addressing sustainability issues related to climate change resiliency and water resources. Brady completed the Kellogg Executive Development Program at Northwestern University and Master of Science and Bachelor of Science degrees in Mechanical Engineering from Drexel University and the University of Rhode Island, respectively.

Christopher Budzynski Budzynski works directly with Exelon Corporation's operating utilities. His current responsibilities include leading the development of strategic growth initiatives and Utility of the Future efforts. Prior to his current role, Budzynski held several management and leadership positions within the Constellation Energy/Exelon Treasury Organization, including Assistant Treasurer at the parent companies and Treasurer of Baltimore Gas and Electric. Before working in the Treasury Organization, Budzynski joined Constellation Energy as a Director of Corporate Strategy, where he was responsible for helping to develop the company's overall strategy direction and supporting various aspects of merger and acquisition activity, including integration work. Prior to joining Constellation Energy/Exelon, Budzynski was Vice President, Utility and Power for Legg Mason, in Baltimore. In this role, Budzynski was a sell-side equity analyst covering the electric, water, and gas industries.

Mark Derry Derry is Vice President of Utility Strategic Planning and is responsible for the oversight of the strategy, policy, and business planning processes for energy providers BGE, ComEd, PECO, and Pepco Holdings. Derry joined ComEd in 1983 as a Staff Accountant and, over the last 30 years, has held positions in various accounting and finance areas, including external financial reporting, regulatory rate case support, corporate accounting, business planning for Exelon Nuclear and EED, and Corporate FP&A. Derry is a CPA and holds a Masters of Accountancy degree from Western Illinois University.

Scott Dupcak Dupcak leads Constellation Technology Ventures (CTV), the venture capital investing arm of Exelon Corporation. In this role, Dupcak oversees the CTV investing strategy, the management of the CTV portfolio, and the commercialization of new energy technologies within the Exelon commercial platform. Prior to CTV, Dupcak was part of the Constellation Strategic Systems and Business Operations group where he led a team responsible for market data management, IT governance and supporting business process improvement projects, new business/growth initiatives, and integration projects. He held several other roles within the Constellation wholesale commercial organization prior to that. Dupcak holds a Master of Business Administration degree from the University of Maryland's Robert H. Smith School of Business and a Bachelor of Science degree in Accounting from Fairfield University.

Kimberly Long Long coordinates and manages Exelon's annual environmental performance and sustainability reporting for environmental resources, including water and biodiversity. Long also coordinates sustainability benchmarking analyses and stewardship certification activities across the company and collaborates with key stakeholders to develop progressive environmental initiatives. Previously, Long served as a FERC License Compliance Manager coordinating environmental compliance commitments such as shoreline and recreation management, fisheries management, and water quality for Exelon's hydroelectric facilities. Long also has experience in environmental permitting associated with wetlands, stormwater, NPDES, and drinking water programs. Prior to joining Exelon, Long served as a watershed manager and aquatic biologist with the Pennsylvania Department of Environmental Protection. Long holds Master of Science and Bachelor of Science degrees in biology from Bucknell University and Millersville University, respectively.

David O'Dowd O'Dowd is a member of ComEd's Corporate Communications department, where he collaborates with a talented team to create awareness and support for ComEd's efforts to transform the way electricity is distributed and used in northern Illinois. O'Dowd began working with ComEd in 2011 after the passage of the historic Energy Infrastructure Modernization Act (EIMA) or Smart Grid Law, which authorized $2.6 billion in grid modernization and smart grid technology investments, including the installation of 4 million smart meters by the end of 2018. O'Dowd's key areas of focus include communications planning and execution for Smart Grid and Technology and the Community of the Future and CLEAR (claims, legal, energy acquisition, and regulatory). David earned Bachelor of Arts degrees in journalism and political science at Indiana University and a Master of Science degree in communications from Northwestern University.

John Slocum Slocum has experience in numerous facets of the legal, regulatory, policy, and commercial aspects of the energy industry. In his role as Manager, State Government Affairs, he is responsible for implementing Exelon's state-level policy, regulatory and legislative advocacy efforts in support of its retail, wholesale, and power generation and development businesses in Illinois and New England. Slocum has guided advocacy initiatives before regulatory organizations and legislative committees in Illinois, New England, and New York on a wide range of energy policy issues, including retail and wholesale competition, competitive wholesale procurement, default service, renewable portfolio standards, carbon policy, and various data and operational issues. On behalf of Exelon, he was intimately involved in the development and passage of the Future Energy Jobs Act (Public Act 99-0906) in Illinois.

Company Profile

Exelon Corporation (NYSE: EXC) is a Fortune 100 energy company with the largest number of utility customers in the United States. Exelon does business in 48 states, the District of Columbia and Canada, and had 2017 revenue of $33.5 billion. Exelon's six utilities deliver electricity and natural gas to approximately 9 million customers in Delaware, the District of Columbia, Illinois, Maryland, New Jersey, and Pennsylvania through its Atlantic City Electric, BGE, ComEd, Delmarva Power, and PECO and Pepco subsidiaries. Exelon is one of the largest competitive US power generators, with more than 35,168 megawatts of nuclear, gas, wind, solar, and hydroelectric generating capacity comprising one of the nation's cleanest and lowest-cost power generation fleets. The company's Constellation business unit provides energy products and services to approximately 2 million residential, public sector, and business customers, including more than two-thirds of the Fortune 100.

References

1. "Learn About Rush Hour Awards" Nest, https://nest.com/support/article/What-is-Rush-Hour-Rewards. Accessed 26 January 2018.
2. "How Project Sunroof Works," Google Project Sunroof, https:/www.google.com/get/sunroof#p=0&spf=1499695123232. Accessed 26 January 2018.
3. St. John, Jeff. "Hawaii Wants 200 MW of Energy Storage for Solar, Wind Grid Challenges." *GTM: A Wood Mackenzie Business*, Greentech Media, 5 May 2014, www.greentechmedia.com/articles/read/hawaii-wants-200mw-of-energy-storage-for-solar-wind-grid-challenges#gs.E30UvpM. Accessed 26 January 2018.
4. St. John, Jeff. "Hawaii Utility's 100% Renewable Energy Plan Gets the Green Light." *GTM: A Wood Mackenzie Business*, Greentech Media, 18 July 2017, www.greentechmedia.com/articles/read/hawaiian-electric-100-renewable-energy-plan-green-light#gs.PigW47Y. Accessed 26 January 2018.
5. Hansen, Louis. "California's push for a 100 percent renewable energy future may hit roadblocks," *The Mercury News*, Bay Area News Group, 13 September 2017, 2:53 p.m., www.mercurynews.com/2017/05/24/california-100-percent-renewable-energy-future-may-hit-roadblocks/. Accessed 26 January 2018.
6. "2016 National Competitiveness Forum, Morning Session, Part 2," 9 December 2016, *C-SPAN.org*, www.c-span.org/video/?419835-3%2F2016-national-competitiveness-forum-morning-session-part-2. Accessed 26 January 2018.
7. Penn, Ivan. "California invested heavily in solar power. Now there's so much that other states are sometimes paid to take it." *Los Angeles Times*, Los Angeles Times, 22 June 2017, www.latimes.com/projects/la-fi-electricity-solar/. Accessed 26 January 2018.
8. "Glossary," U.S. Energy Information Administration, https://www.eia.gov/tools/glossary/index.php?id. Accessed 26 January 2018.

9. Trabish, Herman K. "Inside California's rate restructuring plan and the battle for fixed charges loom over it." Utility Dive, 13 July 2015, https://www.utilitydive.com/news/inside-californias-rate-restructuring-plan-and-the-battle-for-fixed-charge/402117/. Accessed 26 January 2018.
10. Kaufman, K. "Hawaiian utilities ride a wave of innovation, growth and policy debates." *Smart Electric Power Alliance*, 1 October 2015, sepapower.org/knowledge/hawaiian-utilities-ride-a-wave-of-innovation-growth-and-policy-debates/. Accessed 26 January 2018.
11. www.futureenergyjobsact.com/. Accessed 26 January 2018.
12. "NextGrid: Illinois Utility of the Future Study," NextGrid Illinois, www.nextgrid.illinois.gov/. Accessed 26 January 2018.
13. "Notice of Public Conference PC44." *State of Maryland Public Service Commission*, State of Maryland Public Service Commission, 26 Sept. 2016, www.psc.state.md.us/wp-content/uploads/PC-44-Notice-Transforming-Marylands-Electric-Distribution-System.pdf. Accessed 26 January 2018.
14. "MEDSIS Initiative," Public Service Commission of the District of Columbia, https://www.dcpsc.org/Newsroom/HotTopics/MEDSIS-Initiative.aspx. Accessed 26 January 2018.
15. Proterra website, https://www.proterra.com/. Accessed 26 January 2018.
16. Stem website, http://www.stem.com/. Accessed 26 January 2018.
17. Burger, Mark. "Pennsylvania requires solar SRECs to come from within the state." *PV Magazine*, PV Magazine USA, 6 November 2017, pv-magazine-usa.com/2017/11/06/pennsylvania-requires-solar-srecs-to-come-from-within-the-state/. Accessed 26 January 2018.
18. Regular Session 2017–2018, House Bill 1412, Pennsylvania General Assembly, http://www.legis.state.pa.us/cfdocs/billInfo/billInfo.cfm?sYear=2017&sInd=0&body=H&type=B&bn=1412. Accessed 26 January 2018.
19. Stewart, Robert. "Technical Considerations for Transforming the Electric Grid." MADRI Working Group Meeting #45. MADRI Working Group Meeting #45, 7 March 2017, www.madrionline.org/wp-content/uploads/2017/03/MADRI-030717-Final-Presentation-Stewart.pdf. Accessed 26 January 2018.

Chapter 6
A Regulatory Perspective on Transforming the Power Sector

Ann McCabe

Abstract In the forward to EPRI's first sustainable electricity book, World Resources Institute Vice President Janet Ranganathan states that the key sustainability challenge is "how to transform the energy sector into suppliers of clean, dependable, affordable energy services?" (Fox J (ed) Sustainable Electricity, Springer Press, 2016). The power sector is in a state of transition due to growing customer engagement in their energy choices, technological advances such as grid automation, and a focus on decarbonization. These drivers are advancing the growth of distributed energy resources, application of smart grid technologies, and the deployment of renewable and natural gas-fired generation that challenges the future of nuclear generation.

6.1 Introduction

In the forward to EPRI's first sustainable electricity book, World Resources Institute Vice President Janet Ranganathan states that the key sustainability challenge is "how to transform the energy sector into suppliers of clean, dependable, affordable energy services?" [1]. This chapter focuses on the forces driving transformation of the power sector and the resulting sustainability opportunities and challenges, informed by my recent experience as an Illinois State utility regulator from March 2012 until January 2017. I came to the position with over 25 years of energy and environmental policy and regulatory experience. Early in my career during Republican Governor Jim Thompson's administration, I held three successive state positions: a stint as a state budget analyst for several agencies including the Illinois Commerce Commission (ICC), a commissioner's assistant at the ICC for 3 years, and 4 years covering energy and environmental issues for the State of Illinois office in Washington, DC.

In this chapter, I examine several drivers both influencing and influenced by the utility regulatory model and explore the tools available to regulators to enable cost-

A. McCabe (✉)
Independent, Chicago, IL, USA
e-mail: ann@annmccabeinc.com

J. Fox, M. Scott (eds.), *Sustainable Electricity II*,
https://doi.org/10.1007/978-3-319-95696-1_6

effective grid modernization while balancing the affordability needs of the consumer and the financial stability of utilities. At the end of the day, regulators must implement and interpret their state statutory mandate, which in Illinois includes providing adequate, efficient, reliable, least cost, and environmentally safe electricity to the communities they serve [2]. The increased focus on sustainability and the growth in technologies create new challenges and opportunities for regulators in fulfilling their mandates.

The power sector is in a state of transition due to growing customer engagement in their energy choices, technological advances such as grid automation and smart devices, and a focus on decarbonization. For many years, utilities didn't have to worry about change because demand for electricity continued to grow and, with it, revenues. The technology used by utilities to deliver electricity was mature, and utilities had few competitors. Today that has changed, and technological advancements are both transforming utility delivery of electricity and presenting new alternatives. Further, today's electricity demand growth is flat to negative, affecting revenues [3]. The growth in distributed energy resources like customer-sited renewable energy, microgrids, energy storage, and demand response [4], as well as increasing numbers of electric vehicles and the evolution of phones, thermostats, electric meters, and other technologies, has changed the traditional utility business model, created new challenges and opportunities for utilities, and increased consumer expectations of their electricity provider and their home energy management.

Under traditional "cost-of-service" regulation of utilities, the regulator determines a utility's revenue requirement, which reflects the amount to be collected through rates for the utility to recover its operating expenses and capital investment and earn a reasonable rate of return [5]. This method has been employed across all the states in an attempt to balance customer affordability and shareholder returns. As the name implies, cost-of-service focuses on the costs and may not address other qualities desired from modern utility service. The transformation of the power sector is challenging long-held expectations of utilities and regulation and will be shaped in part by decision-makers' ability to develop the tools and markets to cost-effectively accommodate the broader range of available electricity grid resources and market participants.

6.2 The US Regulatory System: An Overview

Before discussing the regulatory tools that can help enable a cost-conscious evolution to a power system of the future, an understanding of today's regulatory environment is needed. State utility regulators are creatures of the legislature: their mandate and authority are set by legislation. They are to regulate in a fair, just, and reasonable fashion to achieve the purpose of their state laws. In many states, regulators are appointed by the governor and confirmed by the state senate; in some states, they are elected. The average length of service of a state utility regulator is about 5 years [6], which can be challenging from a knowledge-transfer perspective for both the regulators and the parties that come before them, including electric power companies, consumer groups, and other stakeholders. In addition to their regular dockets, public utility commissions and their staff can hold public meetings and forums to discuss and learn about important and emerging issues from a variety of stakeholders.

Today state utility regulatory systems fall into two broad categories, vertically integrated or restructured [7]. In the 1990s, a number of states, primarily in the East and the Midwest, began electricity restructuring. These changes followed deregulation of the airline, telecommunications, and natural gas industries. Most restructuring forced the utilities to split competitive generation businesses away from transmission and distribution, giving the customer a choice of electricity supplier, often referred to as "retail choice." Competition among retail electric suppliers can lead to lower electricity prices in restructured states.

As a result of deregulation throughout the 1990s, over 20 states and the District of Columbia have partial or complete competition in power generation and retail energy. In the process of restructuring, wholesale power markets were created, conducted by Regional Transmission Organizations (RTOs) or Independent System Operators (ISOs) (Fig. 6.1). For example, the PJM Interconnection (PJM) coordinates the movement of electricity through all or parts of Delaware, Illinois, Indiana, Kentucky, Maryland, Michigan, New Jersey, North Carolina, Ohio, Pennsylvania, Tennessee, Virginia, West Virginia, and the District of Columbia [8]. Other RTOs and ISOs include California ISO (CAISO), ISO New England (ISO-NE), Midcontinent ISO (MISO), New York ISO (NYISO), Southwest Power Pool (SSP),

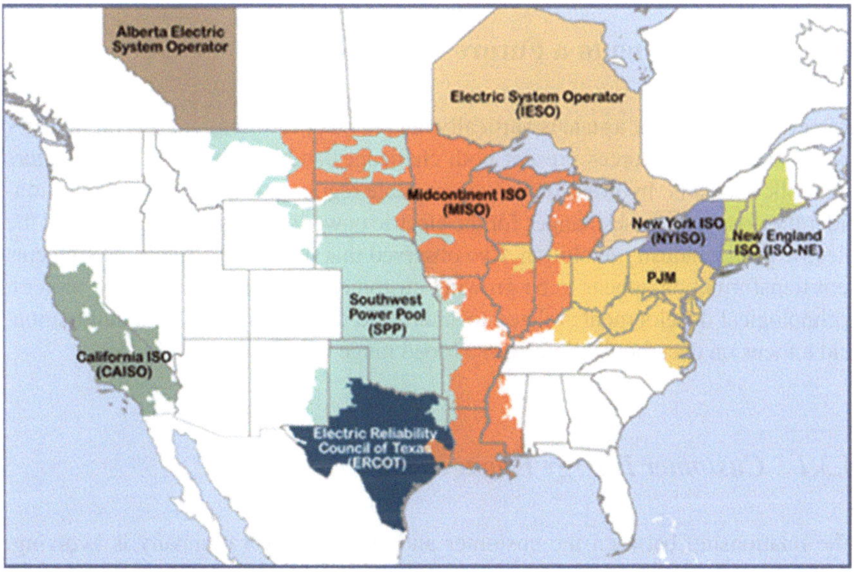

Fig. 6.1 Map of Regional Transmission Organizations (RTOs) and Independent System Operators (ISOs) [10]

and the Electric Reliability Council of Texas (ERCOT). In 2014, the Western Energy Imbalance Market (EIM), a wholesale energy trading market, was launched.

Other states remain vertically integrated, particularly in the southern, central, and northwestern regions of the United States. Vertically integrated utilities are responsible for generation, transmission, and distribution of power to retail customers. Many develop integrated resource plans (IRPs), which are reviewed and approved by state regulators, to identify opportunities for strategic and efficient growth throughout a company's asset value chain. Wholesale sales of electricity are done bilaterally (directly between buyer and seller) rather than through organized markets in some regions (e.g., the South), while dispatch is optimized by system operators in other regions (e.g., MISO, SPP, CAISO).

A state regulator in a vertically integrated state typically has more influence on generation mix than in a restructured state where generation is deregulated, because in deregulated states, the market drives what generation is dispatched. In vertically integrated states, however, where the maintenance, evolution, and dispatch of the generation fleet are controlled through the regulatory-approved IRP, regulators have more ability to influence the way the IRP is designed in order to meet certain desired outcomes. As former Federal Energy Regulatory Commission (FERC) Commissioner, Tony Clark, recently observed about vertical integration, "these [vertically integrated] states can transparently and thoughtfully use electricity policy to pursue a variety of goals" [9]. From these perspectives, vertically integrated or restructured state regulators now consider how they can cost-effectively encourage the evolution of the power industry.

6.3 Transitioning to a Future Power System

Regardless of whether a state is vertically integrated or restructured, several industry trends and external forces present both challenges and opportunities to regulators and policy-makers. In order to evaluate possible regulator responses, one must consider the drivers of these changes for the electric power industry. Having watched the electric power industry evolution, I've observed that three of the forces driving current transformation appear to be growing customer engagement in energy choices, technological developments related to smart grid applications and grid automation, and a focus on decarbonization from a broad range of industry stakeholders.

6.3.1 Customer Energy Engagement

The relationship between the customer and electric power company is evolving. Technology isn't just driving increased electricity consumption from customers; it is also changing the way utilities interact with customers. We now have connected, intelligent devices, loads of data from smart or advanced electric meters, and data analytic capabilities that can increase the grid's efficiency and help customers manage their energy consumption and spending. Customer adoption of technology depends on a number of factors and demographics. Organizations like the Smart Grid Consumer Collaborative analyze the different kinds of customers and to which messages they will respond. Others like Oracle Utilities Opower add information comparing a customer's energy bill to their neighbors. Many customers need to be educated about their options and the capabilities of their smart devices or advanced meters, or they may need to authorize a utility or third party to manage energy demand for them through products and services, with override capability if they want to keep their home warmer or cooler at a specific time. Customer education may include customers understanding that they need to pay more for an electricity grid that enables rooftop solar, utility-scale solar, and wind.

Advanced (or smart) metering infrastructure (AMI)-enabled services and products have been slow to be adopted on the customer side. Some say "the AMI promise isn't there," or "rollout is slower than expected." A recent report lays out ten policy elements to help state public utility commissions (PUCs) address issues with data sharing, including privacy, consumer protections, and technical standards [11]. It recommends utilities adopt Green Button Connect to transfer customer data to third parties, which provides a streamlined customer experience and customer authorization via third-party designs or apps. At present, each state has different rules on data access authorization and authentication. California, Texas, Illinois, New York, and other states far along in implementing advanced meters and rules on data access and Green Button Connect can be helpful guides to "second-mover" states considering advanced metering. Regulators have multiple roles, including protecting customers as they become more engaged in their energy choices as well as enabling technology advancement.

6.3.2 Smart Grid and Grid Automation

The rapid development of technological solutions is also advancing the grid's evolution. Significant advances in grid modernization and automation are reducing some operating costs, increasing reliability and resiliency, and enabling new services and products. The US Department of Energy characterizes the "smart grid" as:

> Modernizing the grid to make it 'smarter' and more resilient through the use of cutting-edge technologies, equipment, and controls that communicate and work together to deliver electricity more reliably and efficiently can greatly reduce the frequency and duration of power outages, reduce storm impacts, and restore service faster when outages occur [12].

The smart grid includes two-way advanced meters that give grid operators and consumers better information and much more data. These allow utilities to identify the location and cause of outages and reroute power so fewer customers are affected in the event of an outage – all without sending a truck. The technology behind the power sector transformation allows the grid to be more resilient to weather and outages. The benefits are many, to the utilities, customers, environment, and the grid.

Benefits of a Smart Grid

In comments from Melissa Lavinson, Vice President of Federal Affairs and Policy and Chief Sustainability Officer, PG&E Corporation, to the Quadrennial Energy Review Task Force on Generating and Delivering Electricity to Meet GHG Targets in 2016, PG&E commented:

"This newer, smarter grid will benefit customers and our environment by:

- Improving electric reliability.
- Facilitating the deployment of new end-use technologies, such as smart appliances, electric vehicles and energy management tools.
- Enhancing efficiency in electricity generation, transmission and distribution.
- Supporting the integration of renewable resources, such as community and large-scale wind and solar and privately owned distributed energy resources" [13].

Beyond grid resilience, the Regulatory Assistance Project states that through enhanced grid communication systems (and data tracking), smarter grids may improve reliability, increase consumer choice, and reduce the economic cost and environmental impact of the utility system. The smart grid is expected to:

- Enable consumers to manage their energy usage and choose the most economically efficient way to meet their energy needs.
- Allow system operators to use automation and a broad array of resources to help maintain delivery system reliability and stability.
- Help utilities to rely on the most economical and environmentally benign resources – generation, demand-side, and storage alternatives – to meet consumer demands [14].

More than 70 million digital smart meters have been installed across the United States. As a result, more than 55% of all American households have the ability to better control their energy use, with further deployment expected in the coming years. Although this technology enables many benefits such as grid hardening and reduced number and duration of outages, challenges remain in ensuring that residential customers see the maximum benefit of smart meters and the wealth of data they generate. Cost-effective rollout and customer data access are just some of the technological considerations that regulators must consider as they evaluate the mechanisms for advancing the future power system.

6.3.3 Decarbonization

Few power sector-related conversations today pass without discussing decarbonization. At both the international and domestic level, decarbonization of the electric power sector has become a focus for a wide array of industry, environmental, and other stakeholders. Climate and energy policy continue to be important to much of the public and to many states and cities, as well as public and private organizations. The investment community and insurance companies increasingly ask companies about the financial risks climate change poses to their assets, infrastructure, and operations. More customers – particularly industrial and commercial customers – want 100% renewable energy to fuel their operations or data centers. The focus on decarbonization will continue, even in the absence of federal policy and regulation.

Increasing efficiency, a greater share of renewable energy, and reducing energy consumption can all result in lower-carbon emissions. Development and deployment of renewable energy for power generation has increased significantly, fueled by policy and regulatory vehicles such as renewable portfolio standards and financial incentives, concerns over energy security, desire for a diversified generation portfolio, continually declining capital costs, and efforts to reduce greenhouse gas emissions.

Despite the United States' June 2017 announcement that it would withdraw from the 2015 Paris Agreement on climate change [15], many government entities, utilities, and companies remain committed to reducing or eliminating fossil fuel-powered generation and increasing energy efficiency and wind and solar generation. In addition to existing climate-oriented groups and coalitions, new state, city, and business groups have pledged to meet the goals of the Paris Agreement. We will continue to see states, cities, companies, non-governmental organizations (NGOs), and other entities pursue lower-carbon goals. With these social drivers, regulators must consider how they can enable electric power companies to meet growing stakeholder demands for choice in energy options and technology while protecting customers from significant rate increases to cover the costs of the required investments.

6.4 Evolution of the Generation Fleet

The drivers of customer engagement, technological advancement, and decarbonization are accelerating a change in the generation portfolio in the United States. From changes in the traditional diversity and nature of generation to the growing proliferation of distributed energy resources, the power system of today looks quite different from the system of the past and will shape the power system of the future. Understanding this evolution helps inform what regulatory tools may be appropriate to help manage potential tradeoffs such changes elicit.

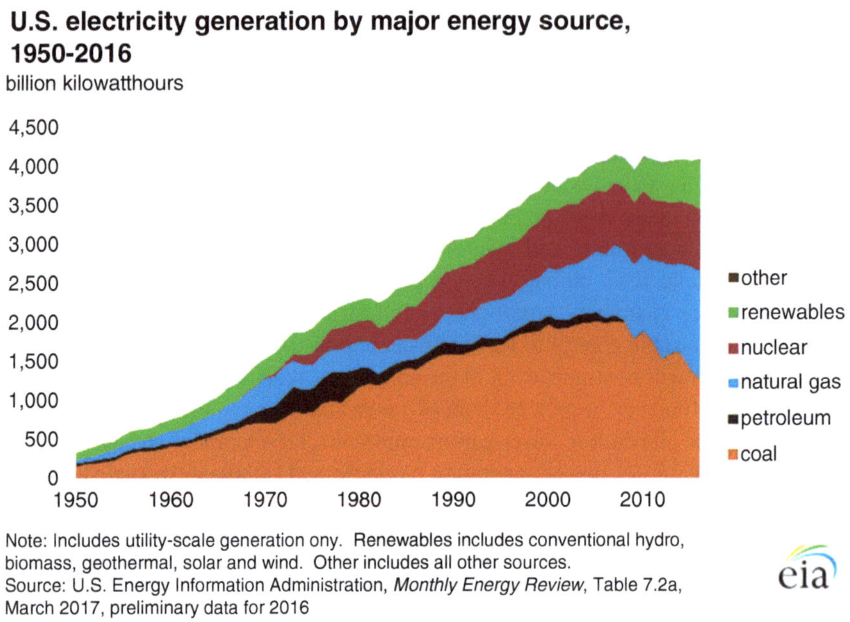

Fig. 6.2 US electricity generation by major energy source, 1950–2016

6.4.1 Evolving Baseload

In 2017, fossil and nuclear generation made up 83% of the United States' total energy consumption (Fig. 6.2), down from about 86% in 2005 [16]. Both wind and solar increased and continue to grow, but renewables remain less than 10% of US power consumption. Some coal and nuclear plants are closing, and the use of natural gas is increasing largely due to its low cost. Companies have additional costs associated with new transmission necessary to connect wind and solar to places of consumption. The limited ability to increase natural gas infrastructure is an issue in places like the Northeast.

The nature as well as the source of generation is changing. Large coal, nuclear, and natural gas-fired power plants have traditionally been called "baseload" plants because they run continuously, providing reliable power, except during maintenance shutdowns. Historically, "baseload" power plants meant generation resources providing relatively low operating cost electricity to meet minimum 24-h/day demand. They were typically fossil fuel or nuclear power plants that take a long time to turn on and off, at a considerable cost. Today, "baseload" has become a function rather than a type of power plant. Now a variety of generation and services provide "baseload services," including renewable energy and demand response that can be turned on and off quickly, as needed. With increased renewables and low-cost natural gas, many plants once providing continuous baseload now need to "ramp" up and down, at times operating at lower capacity and then increasing generation when needed.

Some plants, especially older plants, weren't designed to ramp – a technical, market, and regulatory challenge.

The challenge for policy-makers and regulators is transitioning from a dependence on baseload to a more diverse, distributed generation mix and whether to incentivize certain generation or alternatives to generation until the distribution system and grid can better handle more renewables and storage is more economic.

6.4.2 Nuclear Generation

Traditionally considered a significant contributing source to baseload power, nuclear generation provides 20% of US generation and over 60% of carbon-free generation via 99 nuclear plants (as of 2016) [17]. A typical large nuclear plant provides 1000 MW or more of energy. In 2014, when the polar vortex shut down natural gas and coal plants and disrupted wind and solar, most nuclear kept running, providing power to the hard-hit Northeast [18]. Similarly, the South Texas Nuclear Generating Station near Houston continued to operate during and after Hurricane Harvey in the summer of 2017. These events demonstrate the value of nuclear generation in times of extreme weather.

In recent years, a number of large baseload nuclear plants have closed or announced retirements in both vertically integrated and restructured states. The reasons for closures include nuclear plants' struggle to compete with lower-cost natural gas, significant maintenance and post-Fukushima safety upgrade investments, and higher costs of operating older plants.

Merchant plants owned by independent power producers in restructured states can be at greater risk than plants in vertically integrated states because they can't pass on the costs that utilities can through rate cases (subject to regulatory review and approval). In restructured states that don't regulate generation, merchant plants must compete with low-cost natural gas and renewables in various markets. Entergy, for one, has sold its nuclear plant in the restructured state of New York and has closed or plans to close plants in Vermont, Massachusetts, and Michigan. One question facing policy-makers and regulators is who shoulders the risk: the investors or the customers?

New large nuclear plants being constructed in Georgia and South Carolina have faced challenges due to cost overruns. Santee Cooper and South Carolina Electric & Gas Company spent over $10 billion on two units at the Virgil C. Summer Nuclear Generating Station. South Carolina regulators halted the project in July 2017 due to the bankruptcy of lead contractor Westinghouse [19]. Meanwhile, in December 2017, the Georgia Public Service Commission approved continued construction of the Vogtle nuclear expansion project.

Despite the economic challenges, nuclear energy can be an important player in maintaining fuel diversity and a continuing source of carbon-free emissions and, thus, in the decarbonization of the power sector. When large nuclear plants retire, natural gas has been the cost-effective replacement, resulting in an increase in overall greenhouse gas emissions compared to the previous zero-carbon nuclear generation [20]. In the long term, large-scale renewables and distributed energy resources may help meet demand, so policy-makers need to weigh short- and long-term tradeoffs.

Some argue that existing nuclear plants are at risk because the market (and society) does not recognize the value of their carbon-free emission and reliability attributes. Keeping existing nuclear plants – large and reliable power sources – running could be an important way to prevent carbon emissions from increasing. For viable nuclear power going forward, many call for a solution to the long-term storage of spent nuclear fuel as well.

In the future, scalable small modular nuclear reactors (SMRs) will be the more likely nuclear option given their small size (300 MW and less), scalability or modular nature, lower cost, and greater safety [21]. NuScale, for one, plans a commercial SMR at Idaho National Laboratories for the Utah Associated Municipal Power Systems [22].

6.4.3 Distributed Energy Resources

Distributed energy resources (DERs) include both customer- and utility-owned power systems that are smaller than the traditional power plant. The growth of DER is the result of many drivers and can be seen as the intersection between trends around customer engagement in energy choices and an interest in decarbonization. As noted in a Utility Dive article, "many of the key tensions regulators now face involve reconciling traditional 'cost-plus' [cost-of-service] regulation with climate goals and the complicated economics of emerging distributed energy resource (DER) technologies" [23].

In the last decade, consumers seeking to reduce their energy bills and environmental footprint, and be more energy independent, have been installing rooftop solar and other DER. Prices have fallen to the extent that renewable energy is cost-competitive with the electricity provided by gas- and coal-fired plants in much of the country, which has continued to drive the deployment of these customer-sited generation sources. According to a Lawrence Berkeley National Lab report, the cost of a residential solar system has fallen to around 25–30 cents/kWh. With additional federal and state subsidies and tax benefits, the cost falls to 15 cents/kWh. "Generally speaking, if the retail rate for electricity is less than that, solar is a poor investment; if it is more than that, rooftop solar is a good investment" [38].

Like any technology, DERs have both pros and cons. Their broad system benefits can include an increase in grid resiliency due to deployment during peak energy demand times and power outages due to storms or equipment failure. In the years ahead, however, utilities and policy-makers will need to address challenges posed by DER, including reliability and safety concerns associated with two-way power as distributed generation like wind and solar feed power back into the utility system. Utilities must modify the grid to handle two-way power flows and prevent distribution circuits designed for one-way flows from getting overloaded and threatening reliability. California, Hawaii, Arizona, New Jersey, and other states [39] have taken the lead on solar integration and may serve as examples for others in the years ahead.

Often discussed in tandem with the growth of customer-sited renewable generation is the evolution of battery technology. Batteries can enable storage of renewable energy and voltage optimization to avoid investment and help meet peak demand times, especially in remote areas or overburdened electricity distribution nodes. Having the ability to store daytime solar increases the grid's reliability and resiliency.

"Behind-the-meter" (BTM) refers to storage, wind, rooftop solar, and other energy generation on the customer side of the meter. The BTM energy storage market is small now but could be 50% or 2 GW of the US energy storage market by 2021 [38]. This market is driven by improved system economics, changes to net metering policies and utility rate structures, increasing viability of incentives to customers who curtail or shift energy demand to nonpeak times, and increasing interest in reliability and resiliency [38]. Combining storage with solar power allows customers to take advantage of declining solar power costs and store solar-generated power so they can shift their load to reduce the amount of expensive peak power purchased from their utility.

Hawaii is an example of the changes driving the adoption of energy storage in combination with solar and setting the stage for the BTM storage market. The Kauai Island Electric Cooperative and AES are combining a 28-MW solar farm with a 200-MW, 100-MWh battery system to enhance the dispatch capability of solar power to the grid. The cooperative has agreed to pay $0.11/kWh for the output from the solar-plus-storage system, which is below the island's current cost of baseload power and represents a 30% drop in costs from a purchased power agreement deal made 2 years ago for $0.145/kWh [38]. This type of storage collaboration may help Hawaii achieve its goal to be 100% renewably powered by 2045.

Storage isn't just for solar installations. In 2009 S&C Electric and Xcel Energy built the first wind-to-battery storage project, Minwind, in Beaver Creek, Minnesota, that can power 500 homes for 7 hours [40]. S&C provided the storage management system, including a 1-MW sodium-sulfur battery system and a backup diesel generator to heat the batteries.

The intermittency of wind and solar will likely spur investment in battery storage where appropriate and feasible. In the next decade or so, we will see more utility-scale (large) storage and renewables as well as aggregated small-scale storage. State regulators will increasingly need to determine cost recovery for battery storage projects, whether stand-alone or part of a microgrid. They should consider energy storage options in transmission, generation, and distribution planning.

With the changing landscape of power generation associated with evolving perspectives of baseload, advancement of renewable technology and deployment, and continued adoption of DER, regulators are faced with new questions around how to enable power system transformation while ensuring affordability for customers, financial stability for utilities, and returns for shareholders.

6.5 Changing Regulatory Approaches for a Changing Power System

A regulator can utilize a number of tools to change rate design and regulatory models to better meet changing electricity grid dynamics and future needs cost-effectively. The Rocky Mountain Institute (RMI) has observed, "*Market design can be improved.* U.S. electricity markets were designed for an era when large, central power plants competed only with each other… [M]any observers agree that there is a pressing need to reexamine how we define products and compensate market players to ensure that our market rules reflect this paradigm shift in the costs and capabilities of nontraditional resources" [24].

Regulators are faced with the need to "encourage innovation while protecting consumers" from inappropriate utility investments [25]. Under traditional cost-of-service regulation, utilities often pursue capital expenditures on which they can earn a return. However, regulators have a number of options in modifying the regulatory model, depending on their legislative authority, through which they can help drive the US grid toward a least-cost, reliable, and resilient future [26]. These modifications include decoupling, encouraging non-wires alternatives, implementing performance-based regulation, and establishing multi-year rate plans.

6.5.1 Decoupling and Non-Wires Alternatives

Decoupling is a tool that many states – but not all – have enacted as a way to remove a disincentive for utilities to embrace energy efficiency or other measures that reduce consumer consumption. Decoupling separates a company's profits from the sale of electricity, essentially helping the utility stay financially viable if energy consumption decreases. Rate increases or decreases determined through a regulatory rate case can maintain utilities' revenues.

The need for utility infrastructure and electricity transmission investment in the United States is several trillion dollars, both for upgrades/modernization and new investment [27]. An unintended consequence of decoupling can be the over-reliance on system investment by transmission and distribution companies to meet growing demand. With this in mind, regulators have begun to push for alternatives such as non-wires or non-transmission projects that defer or eliminate the need for expensive investment. This is a shift for electricity distribution companies. Non-wires alternatives can defer or replace investment in wires, conductors and other distribution infrastructure.

For example, in New York, Consolidated Edison (ConEd) is using non-wires alternatives to defer a $1.2 billion substation upgrade through demand management resources such as demand response, solar, efficiency, and storage. In August 2016, ConEd held a successful auction for demand response capacity for Brooklyn Queens: it will address peak hours' demand by offering incentives to those who reduce demand and/or build storage [28]. An award-winning microgrid project at a

Brooklyn apartment complex combines a 400-kW solar photovoltaic (PV) system and a 400-kW fuel cell supported by a lithium-ion battery storage system. The system enables the building to use solar energy stored during the day in the evening hours. It gives the apartment complex resiliency, revenue and savings through demand charge reduction, capacity relief, and other grid services [29].

Given the project's success to date, it was extended in the summer of 2017, and ConEd plans to expand the "non-wires solutions" program [30]. Matthew Ketschke, ConEd senior vice president, said:

> "There are going to be a lot of times when traditional infrastructure is the lowest cost, best alternative, and we will build traditional infrastructure... And there will be other times when customer-side solutions and a non-wires solution will be the best solution, and we'll go that way [30]."

We will likely see more creative, least-cost solutions, including "non-wires alternatives" or "non-generation" solutions in the coming years.

6.5.2 Performance-Based Regulation (PBR)

Performance-based regulation (PBR) gives utilities a chance to increase revenues by meeting metrics that can include energy efficiency gains and reduced outages. In performance-based regulation, utilities are incentivized to cut operating costs to earn more revenue. Changing the way utilities make money and giving them an incentive to operate the distribution system in a way that minimizes costs becomes more important in a future with high distributed energy resources. PBR focuses on achievement of outcomes rather than how the utility achieves them. To succeed, PBR must create good, outcome-based incentives through clear, measurable metrics, remove bad incentives, be transparent, learn from experience, and make modifications as necessary [31].

As part of its Reforming the Energy Vision (REV) process, New York is trying to streamline the integration of DER and to measure utility performance through a proposed PBR mechanism known as an "Earning Adjustment Mechanism" (EAM). An EAM can adjust utility earnings upward or downward depending on how it meets certain performance metrics. For example, a utility could be rewarded for getting distributed resources online faster. As such, the REV is moving away from traditional ratemaking to compensate utilities for the cost of new services.

California also has been pursuing PBR standards, with a "Distribution Planning Advisory Group" requesting utilities facing large system upgrades to solicit proposals to meet distribution system needs through DER. PBR has been utilized in Ohio through the PowerForward program. Minnesota's electricity in the twenty-first century (E21), an initiative to develop more customer-centric rates and a sustainable framework for utility regulation, is in its third phase and looking at performance-based regulation [32]. Rhode Island, Washington DC, and other jurisdictions have begun looking at PBR to enhance their regulatory approach to enabling the power system of the future [33]. In 2017 Illinois began a NextGrid proceeding [34].

Each state gives its flavor to the debate, but the aim is similar: devise ways to compensate utilities for new energy services, like efficiency or DER deployment, that do not fit into the traditional utility ratemaking model. The incentives for successfully doing so include the utility recovering the cost of DER acquisition, a return on equity adder, and sharing the savings with customers.

6.5.3 Multi-year Rate Plans

Any company that reports to a state regulating body submits a rate case to set pricing for customers. Multi-year rate plans can have pre-approved rate increases tied to various factors or changes in costs. Plans cover all aspects of an electric power company's activities, including programs to help customers manage electric bills, increase reliability, and enhance energy efficiency. The benefits of a multi-year rate plan include giving the utility rate certainty for several years, reducing regulatory costs since the utility has fewer proceedings before the state utility regulators, and, in the best scenario, providing lower, more stable rates for customers.

6.5.4 Subsidy Case Study: Illinois and New York Zero Emission Credits for Nuclear

With several tools available to regulators, two restructured states, Illinois and New York, approved subsidies in 2016 to keep specific nuclear plants from retiring. As discussed in Sect. 4.2, nuclear plants provide significant power: most plants are 1000 MW or larger. In recent years, some nuclear plants have had difficulty competing with low-price natural gas and/or face significant capital and operating expenses, especially if they are older. In 2016, Illinois and New York instituted Zero-Emission Credits (ZECs), a financial mechanism yielding revenues that will help preserve specific nuclear plants and the related jobs and zero-carbon emissions. The tradeoffs include increased rates and potential wholesale market distortions. As Resources for the Future points out in a 2017 report:

"Policymakers, regulators, system operators, or publicly owned utilities in any part of the world may wish to prevent unprofitable generators from retiring for any number of reasons. They may wish to preserve the jobs of coal miners or plant workers, they may prefer not to make the capital investment necessary for replacement capacity, or they may consider the plants to be important for reliability, for resilience, or for limiting harmful emissions." [35]

Illinois has the most nuclear generation of any US state, with 11 units at 6 plants, providing about 50% of the state's electricity. In December 2016, the Illinois legislature passed SB 2814, or the Future Energy Jobs Act (FEJA), which creates ZECs to help preserve Exelon's Clinton and Quad Cities nuclear plants and their combined 1500 jobs. The Illinois Act uses a social cost of carbon to determine the ZEC's price, and the Illinois Power Agency will procure the ZECs. For the first 6 years, the subsidy will be $16.50 per MWh. Prior to the legislation, both PJM and MISO (the Regional Transmission Organization and Independent System Operator for Illinois) said closure of the plants did not present resource adequacy issues for the regions.

The projected cost to Illinois ratepayers is up to $235 million a year for 10 years, or $2.35 billion total. Exelon, the owner of the plants, can bank the credits; it is not regulated by the state since Illinois is restructured with no regulatory oversight over generation. Since Illinois exports 30–40% of its power generation, the ZECs force Illinois ratepayers to subsidize power sales to neighboring states to the east [36].

On August 1, 2016, the New York Public Service Commission approved a proposal that included Zero-Emission Credits (ZECs) for three upstate nuclear plants – R. E. Ginna, James A. FitzPatrick, and Nine Mile Point. The policy, effective in 2017, compensates upstate nuclear plants for every MWh of carbon-free electricity generated and is projected to cost around $7.6 billion over a 12-year period, an estimated $2 increase on the average monthly bill. The first 2 years of the program are expected to cost $965 million when upstate nuclear plants will receive ZEC compensation at a rate of $17.48 per MWh, based on the social cost of carbon and the avoided carbon emissions the plants represent [37].

These subsidies have economic and policy impacts on several levels. While the subsidies, which will last about 10 years, mean increased customer costs in both Illinois and New York, they protect nuclear jobs and provide certainty to the utilities and the investment community around this zero-carbon source of power. At the wholesale level, they compete with or undermine new generation options and may suppress prices (profits) for existing generation. Ideally, from my policy perspective, ZECs should be available to any zero-carbon source rather than individual plants. Then such sources could compete with each other in the procurement auctions and the market.

For a regulator or legislator, tradeoffs must be considered and managed – in this case, how to balance a growing desire to decarbonize the grid and to preserve specific nuclear units and their local jobs while simultaneously keeping costs to customers affordable. The ZEC subsidy decision is an example of how Illinois legislators and New York regulators chose to balance customer affordability, shareholder needs, and societal drivers in deciding who shoulders risk in preventing certain generation sources from retiring.

6.6 Conclusion

In the next few decades, we will see continued advances in grid automation technology, growth in renewables with an accompanying need for grid flexibility, increased viability of storage, and a proliferation of customer-sited resources. These advances will lead to regulatory change in some places and more ways for customers to manage their energy demand and costs and will require distribution and transmission system operators to be more nimble and flexible. The tradeoff may be higher electricity prices ($/kWh) to cover the costs of new and improved grid and transmission and, in some cases, to preserve or ensure a mix of power generation sources. It is an open question whether investments in a smarter, cleaner, and more flexible grid will result in higher customer costs, or whether these new capabilities, new energy management products and services, and behind-the-meter resources can be leveraged in a way that results in lower costs and greater overall benefits for customers.

Advances in technology, software, and distributed energy resources can help increase the grid's resiliency, enable energy demand management that minimize greenhouse gas emissions, and increase efficiency. In the absence of federal climate change policy, a number of states, cities, companies, and organizations will continue to reduce their greenhouse gas emissions and demand greener power and more energy efficiency. As a former regulator and someone still active in the energy and carbon conversation, sustainable electricity to me means power generation, distributed energy resources, and a grid that ensures the reliability, resiliency, and affordability of electricity in a resource-efficient and environmentally protective manner for the long term. As this chapter discusses, regulators have a number of tools to enable this technological advancement, or transformation of the power sector, while balancing the interests of utilities, customers, service providers, and others.

Author Profile

Ann McCabe McCabe is a consultant and interim executive director of The Climate Registry, which helps companies, state agencies, universities, and others measure, report, and verify their greenhouse gas emissions. For 9 years, The Climate Registry has co-led a delegation of state officials to the UN's Conference of the Parties (COP) on climate change. From March 2012 to January 2017, as an Illinois public utility commissioner, McCabe was president of the Mid-America Regulatory Conference and the Organization of PJM States, Inc. boards and chaired the National Association of Regulatory Utility Commissioners' Nuclear Issues Subcommittee. McCabe previously consulted to industry, trade associations, and nonprofits; managed environmental regulatory issues for BP and Amoco; and served in the Illinois Washington DC office. She has a master's degree in public policy from the University of Chicago and a bachelor's degree in political science from Williams College and is a regular speaker on grid modernization, energy efficiency, and nuclear power.

References

1. Fox, J., Editor. *Sustainable Electricity*, May 2016. Springer Press.
2. Illinois General Assembly, Illinois Public Utility Act (220 ILCS 5/1–102), http://www.ilga.gov/legislation/ilcs/ilcs5.asp?ActID=1277&ChapterID=23. Accessed 7 March 2018.
3. U.S. Energy Information Administration. "Today in Energy" 29 April 2015, https://www.eia.gov/todayinenergy/detail.php?id=21012. Accessed 7 March 2018.
4. "Integration of Distributed Resources" Electric Power Research Institute, https://membercenter.epri.com/programs/067418. Accessed 11 May 2018.
5. "Tariff Development: The Basic Ratemaking Process, https://pubs.naruc.org/pub.cfm?id=538E730E-2354-D714-51A6-5B621A9534CB. Accessed 7 March 2018.

6. Data compiled and provided by Janice Beecher, Director, Institute of Public Utilities, Michigan State University, Commissioner Demographics 2018, ipu.msu.edu, April 7, 2018.
7. Map of Deregulated Energy States and Markets (Updated 2017), Electric Choice, https://www.electricchoice.com/map-deregulated-energy-markets/. Accessed 7 March 2018.
8. "Territory Served," PJM, http://www.pjm.com/about-pjm/who-we-are/territory-served.aspx. Accessed 7 March 2018.
9. Clark, Tony et al. "Regulation and Markets: Ideas for Solving the Identity Crisis," July 2017, p. 10., http://www.wbklaw.com/uploads/file/Articles-%20News/2017%20articles%20publications/Market%20Identity%20Crisis%20Final%20(7-14-17).pdf. Accessed 7 March 2018.
10. http://www.theenergycollective.com/aqgilbert/2322195/us-electricity-system-15-maps. Accessed 26 May 2018.
11. "AEMA and Mission:data Release Policy Guidebook on Consumer Energy Data," Advanced Energy Management Alliance, http://aem-alliance.org/aema-missiondata-release-policy-guidebook-consumer-energy-data/. Accessed 7 March 2018.
12. "Grid Modernization and the Smart Grid," U.S. Department of Energy Office of Electricity Delivery & Energy Reliability, https://energy.gov/oe/activities/technology-development/grid-modernization-and-smart-grid. Accessed 7 March 2018.
13. "PG&E Corporation before the Quadrennial Energy Review Task Force on Generating and Delivering Electricity to Meet GHG Targets," May 10, 2016, https://www.energy.gov/sites/prod/files/2016/05/f31/Panel%203%20Remarks%20by%20Melissa%20Lavinson%2C%20Vice%20President%2C%20Federal%20Affairs%20and%20Policy%20and%20Chief%20Sustainability%20Officer%2C%20Pacific%20Gas%20and%20Electric%20%28PG%26E%29.pdf. Accessed 7 March 2018.
14. Electricity Regulation in the U.S.: A Guide (Second Edition). Regulatory Assistance Project, http://www.raponline.org/wp-content/uploads/2016/07/rap-lazar-electricity-regulation-US-june-2016.pdf, and "Inside the New Guidebook for Electric Utility Regulation," Utility Dive, 22 September 2016, http://www.utilitydive.com/news/inside-the-new-guidebook-for-electric-utility-regulation/426597/. Accessed 7 March 2018.
15. "The Paris Agreement," United Nations, http://unfccc.int/paris_agreement/items/9485.php. Accessed 7 March 2018.
16. "What is U.S. electricity generation by revenue source?" U.S. Energy Information Administration, https://www.eia.gov/tools/faqs/faq.php?id=427&t=3 and "U.S. energy consumption rose slightly in 2016 despite a significant decline in coal use," U.S. Energy Information Administration, 5 April 5 2017, https://www.eia.gov/todayinenergy/detail.php?id=30652. Accessed 3 April 2018.
17. "What is U.S. generation by fuel source?" U.S. Energy Information Administration, https://www.eia.gov/tools/faqs/faq.php?id=427&t=3. Accessed 7 March 2018.
18. "Polar Vortex Review," NERC, September 2014, https://www.nerc.com/pa/rrm/January%202014%20Polar%20Vortex%20Review/Polar_Vortex_Review_29_Sept_2014_Final.pdf. Accessed 2 March 2018.
19. "Prosecutors Subpoenaing Documents in Failed Nuke Project," StarTribune, 22 September 2017, http://www.startribune.com/feds-appear-to-be-probing-possible-nuclear-project-fraud/446847583/. Accessed 3 March 2018.
20. "Atomic – How the Nuclear Power Meltdown will Help U.S. Natural Gas Producers," Housely Carr, 15 June 2017, https://rbnenergy.com/atomic-how-the-nuclear-power-meltdown-will-help-us-natural-gas-producers and https://www.eia.gov/naturalgas/weekly/archivenew_ngwu/2017/09_28/. Accessed 3 March 2018.
21. "Advanced Small Modular Reactors (SMRs)," U.S. Department of Energy Office of Nuclear Energy, https://www.energy.gov/ne/nuclear-reactor-technologies/small-modular-nuclear-reactors and "With New Reactors, a Better World Awaits," NEI, https://www.nei.org/Issues-Policy/New-Nuclear-Energy-Facilities/Small-Reactor-Designs. Accessed 3 February 2018.
22. "NuScale's Small Modular Nuclear Reactor Keeps Moving Forward," Forbes, 16 March 2016, https://www.forbes.com/sites/jamesconca/2017/03/16/nuscales-small-modular-nuclear-reactor-keeps-moving-forward/#40741a925456 and "Utah Associated Municipal Power Systems

(UAMPS) Carbon Free Power Project (CFPP) Progresses," NuScale Power, http://www.nuscalepower.com/our-technology/technology-validation/program-win/uamps. Accessed 3 February 2018.
23. "Electricity Regulation in the U.S.: A Guide (Second Edition)," Regulatory Assistance Project, http://www.raponline.org/wp-content/uploads/2016/07/rap-lazar-electricity-regulation-US-june-2016.pdf, and "Inside the New Guidebook for Electric Utility Regulation," Utility Dive, 22 September 2016, http://www.utilitydive.com/news/inside-the-new-guidebook-for-electric-utility-regulation/426597/. Accessed 3 March 2018.
24. Dyson, Mark. "Changes in the Power Sector are an Opportunity, Not a Threat," RMI Outlet, 24 August 2017, https://rmi.org/news/changes-power-sector-opportunity-not-threat/. Accessed 7 March 2018.
25. "Inside the New Guidebook for Electric Utility Regulation," Utility Dive, 22 September 2016, http://www.utilitydive.com/news/inside-the-new-guidebook-for-electric-utility-regulation/426597/. Accessed 7 March 2018.
26. "Electricity Regulation in the U.S.: A Guide (Second Edition)," Regulatory Assistance Project, http://www.raponline.org/wp-content/uploads/2016/07/rap-lazar-electricity-regulation-US-june-2016.pdf. Accessed 7 March 2018.
27. Paraskova, Tsvetana. "Replacing the US electric grid could cost $5 trillion," OilPrice.com, 25 March 2017, http://www.businessinsider.com/replacing-the-us-electric-grid-could-cost-5-trillion-2017-3. Accessed 7 March 2018.
28. Bade, Gavin. "ConEd awards 22 MW of demand response contracts in Brooklyn-Queens project," Utility Dive, 8 August 2016, http://www.utilitydive.com/news/coned-awards-22-mw-of-demand-response-contracts-in-brooklyn-queens-project/424034/. Accessed 7 March 2018.
29. "Award-Winning Microgrid in Brooklyn "REVolutionizes" the Electricity Market," CleanTechnica, 30 August 2017, https://cleantechnica.com/2017/08/30/award-winning-microgrid-in-brooklyn-revolutionizes-the-electricity-market/. Accessed 7 March 2018.
30. Walton, Robert. "Straight Outta BQDM: Consolidated Edison looks to expand its non-wires approach," Utility Dive, 19 July 2017, http://www.utilitydive.com/news/straight-outta-bqdm-consolidated-edison-looks-to-expand-its-non-wires-appr/447433/. Accessed 7 March 2018.
31. Littell, David, et al., "Next Generation Performance-Based Regulation: Emphasizing Utility Performance to Unleash Power Sector," RAP, 12 September 2017, http://www.raponline.org/knowledge-center/next-generation-performance-based-regulation-emphasizing-utility-performance-unleash-power-sector-innovation/. Accessed 7 March 2018.
32. "About the e21 Initiative," e21 Initiative, http://www.betterenergy.org/projects/e21-initiative. Accessed 7 March 2018.
33. Bade, Gavin. "10 trends shaping the power sector in 2018," Utility Dive, 22 January 2108, https://www.utilitydive.com/news/10-trends-shaping-the-power-sector-in-2018/515235/. Accessed 7 March 2018.
34. "Next Grid: Illinois Utility of the Future Study," NextGrid Illinois, https://nextgrid.illinois.gov/. Accessed 7 March 2018.
35. Shawhan, Daniel and Picciano, Paul. Costs and Benefits of Saving Unprofitable Generators: A Simulation Case Study for US Coal and Nuclear Power Plants, RFF WP 17–22, Resources for the Future, November 2017, http://www.rff.org/files/document/file/RFF-WP-17-22.pdf. Accessed 7 March 2018.
36. "Illinois Profile Overview," U.S. Energy Information Administration, https://www.eia.gov/state/?sid=IL. Accessed 7 March 2018.
37. "State Options to Keep Nuclear in the Energy Mix," National Conference of State Legislatures, http://www.ncsl.org/research/energy/state-options-to-keep-nuclear-in-the-energy-mix.aspx and "REC and ZEC Purchasers," New York State Clean Energy Standard, https://www.nyserda.ny.gov/All-Programs/Programs/Clean-Energy-Standard/REC-and-ZEC-Purchasers. Accessed 7 March 2018.

38. Maloney, Peter. "How behind-the-meter storage could make up 50% of the U.S. market by 2021, Utility Dive, 31 January 2017, http://www.utilitydive.com/news/how-behind-the-meter-storage-could-make-up-50-of-the-us-market-by-2021/434882/. Accessed 3 March 2018.
39. "Top 10 Solar States," Solar Energy Industries Association, http://www.seia.org/research-resources/top-10-solar-states. Accessed 3 March 2018.
40. "S&C Electric Company: First Wind Energy Storage Application in the U.S.," Energy Storage Association, http://energystorage.org/energy-storage/case-studies/sc-electric-company-first-wind-energy-storage-application-us. Accessed 7 March 2018.

Chapter 7
The Third "E": LADWP's Commitment to Equity Ensures First-Rate Service to All

Nancy Sutley, William Funderburk, John Chen, David Jacot, and Maria Sison-Roces

Abstract How does a utility with an annual budget of nearly $5 billion, providing services to 1.5 million electric customers and over 680,000 water customers and covering an area of 465 square miles, make sure that it is serving its customers equitably? That is the question that the Los Angeles Department of Water and Power (LADWP) Board of Water and Power Commissioners set out to answer when it adopted the Equity Metrics Data Initiative (EMDI) in August of 2016. In this chapter, we will share our very own journey into sustainability as a municipal utility, while highlighting the importance of equity as the ubiquitous core value of our business practices that drives our commitment to sustainability and customer service. Our focus on equity also necessitates some tradeoffs, and these, too, are discussed.

7.1 Introduction

As a customer base, it's hard to imagine a city more diverse than Los Angeles. LA is a melting pot where over 185 languages [1] are spoken by over 4 million people. It is a unique blend of cultures and economic conditions distributed within 15 council districts and represented by 96 neighborhood councils. With the advent of demand-side management, decarbonization, distributed energy resources, and electrification of the transportation sector, things are getting "quite electric" in the City of Angels. At the same time, LA is a values-driven city. These values are clearly defined in LA's Sustainability pLAn, where the city's goals are based on our three defined pillars of sustainability – environment, economy, and equity.

Traditionally referred to as the "social" pillar of sustainability, the Los Angeles Department of Water and Power (LADWP) has focused on the vital "E" for equity, defined as the quality of being fair and impartial, which is central to our customer-centric mission. It is a core principle in Los Angeles and is consistent with the city

N. Sutley · W. Funderburk · J. Chen · D. Jacot · M. Sison-Roces (✉)
Los Angeles Department of Water & Power, Los Angeles, CA, USA
e-mail: Maria.Sison-Roces@ladwp.com

J. Fox, M. Scott (eds.), *Sustainable Electricity II*,
https://doi.org/10.1007/978-3-319-95696-1_7

leadership's policies on equal employment, workforce, business opportunity, gender equity, and affordable housing, as well as direction on equitable clean energy solutions and senior and low-income customer programs. As the city's municipal utility, LADWP has committed to vastly enhance customer engagement and service by ensuring that its services and operations reach all customers fairly.

Equity is integrated into the way LADWP manages its operations, makes decisions, deploys its resources, provides services to its customers, and engages local businesses and the collaborative economy. Through LADWP's Equity Metrics Data Initiative (EMDI) that will be enhanced in the years ahead, indicators of equity are monitored, tracked, and utilized to ensure that water and electricity are part of the palette of tools for creating a just society.

As a public agency dealing with mostly public funds, LADWP has a responsibility to serve all its customers fairly and equitably. The Board of Water and Power Commissioners, comprised of five appointed members from the community, makes multimillion-dollar decisions on behalf of Los Angeles. They wanted to know: What are the impacts of our decisions? How can we be sure that LADWP's resources are being used in the best way possible? Looking to the future, how do we serve our customers better and integrate equity into our programs and services?

For LADWP, equity is not just about low-income and disadvantaged customers. It is about serving all customer segments fairly in all geographic areas, whether it's a single or multifamily home; a small, medium, or large business; or an industrial customer, regardless of zip code. As the economy ebbs and flows and as communities experience changes in population, diversity, and demographics, equity metrics can objectively reflect the communities that are availing of LADWP's services. This valuable insight helps inform LADWP of the reach and impact of its programs and offerings and helps us determine when to refocus resources and recalibrate outreach to further connect with intended communities.

7.2 The Case for Equity Metrics

Los Angeles is a regionally, culturally, and economically diverse city, with a large share of disadvantaged and underserved customer segments. Regionally, the city includes suburban residential and manufacturing areas, concentrated business districts, tourism and entertainment districts, and beach communities, all falling within 96 distinct neighborhoods (Fig. 7.1).

LADWP's customer base also varies widely. Housing demographics are mixed, with home ownership averaging 36% city-wide but ranging from under 20% in some areas of the city to over 50% in others [2]. Culturally, nearly 40% of residents are foreign born [3], with over 185 languages spoken at home [4]. Educationally, adults with a high school education or higher make up an average of 76% of the population [5], with the percentage ranging from a low of 25% in some areas to a high of 97% in others [6]. While the median household income in 2016 was $51,538 for the city [3] as a whole, poverty levels cover the full range of less than 10% to

Fig. 7.1 Diversity in Los Angeles

greater than 90% for different areas of the city [7]. Businesses range from mom and pop corner stores in areas like Chinatown to high-tech campuses such as the La Kretz Innovation Campus in the Arts District.

From the customer viewpoint, many factors affect their needs and how they participate in customer incentive programs. Economic diversity affects purchasing decisions and how energy and water are used. Whether someone lives in their own home or rents has a large impact on the applicability of many customer programs. Many renters are unable to implement the same type of solutions as property owners (e.g., solar installation). Cultural diversity affects the way people learn about customer programs. Some customers are easily reached through the Internet and social media, while others rely on trusted advisors and personal contact. Extra effort is required to reach across cultural and language barriers.

Each of these diversity factors underscores the need for different approaches for different customers. Equity means treating everyone fairly versus equality, which means treating everyone the same (Fig. 7.2). We know that things are not equally needed or valued by our diverse customer base. This means that we need to understand and respond to what is equitable and fair based on the changing needs of our unique customers. Nowhere is that more applicable than with the variety of approaches that are needed to ensure equity in providing services and programs in the LADWP service area.

So, what is fairness? Who determines what is fair? Defining what is fair is part of the process. Fairness depends on the observer's point of view. Customers, the utility, community leaders, elected officials, and other stakeholders may have different perceptions of what is fair. The path to determining equitable solutions involves engaging all these groups.[1] LADWP established EMDI as a tool to better track,

[1] For more information on environmental justice and sustainability, see Refs. [8, 9].

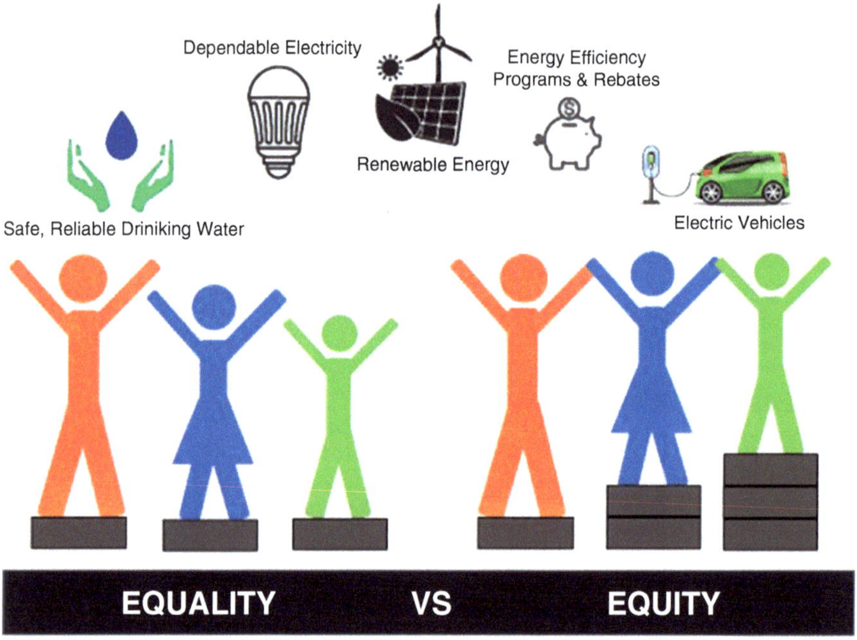

Fig. 7.2 Equity vs equality

measure, and report on how our programs are provided to all customers and residents of Los Angeles.

7.3 Equity Metrics Data Initiative (EMDI)

The Equity Metrics Data Initiative, introduced by Board Vice President William Funderburk, adds another dimension to the accountability and transparency equation. It is an effective way to better serve and build trust among LADWP's constituents. It is a data-driven framework that is used to assess how well programs, services, and resources are distributed throughout the city. Data can be examined both geographically and demographically to see whether and where any disparities exist. Data collection and analysis through EMDI provides important information about LADWP's services and operations. It also provides a tool for decision-makers to ensure that all customers are reached with fairness and equity. The initiative also enables the department to weave equity throughout the organization and embed it as a cornerstone of LADWP management and Board's best practices (Fig. 7.3).

EMDI is a way to tell an objective story in a visual manner. Utilities tend to have a lot of data, yet from the customer's perspective, sometimes spreadsheets and pie charts don't tell the whole story. As the adage goes, one can be data rich and information poor. Tracking this data helps LADWP make informed decisions and man-

William Funderburk, Vice President, LADWP Board of Water and Power Commissioners

"The Equity Metrics Data Initiative is not just about equity, it's about our culture here at LADWP. As a Commissioner, I've come to see that providing public water and power is about balancing the needs of the community with the needs of all ratepayers by providing water and power at the most competitive cost and making sure that we deploy our resources equitably. EMDI empowers our customers to hold us accountable as a Commission and make sure that we live up to our equity goals. Equity metrics helps to engage the community and informs us as policy makers with data that is based on facts rather than perception so that we can do the right thing by the ratepayers."

Fig. 7.3 Equity matters, the role of public utilities. (Personal communication, 2 October 2017.)

age tradeoffs by seeing the data mapped to actual neighborhoods. It also sends a message that LADWP pays attention to the customer who may wonder, "When my lights go out in South LA, will I receive the same level of service as if I lived on the Westside?"

With input from key stakeholders, staff identified 50 equity metrics [10], from which 15 were selected for EMDI's initial focus in four core categories that are the most relevant to multiple aspects of LADWP's operations: water and power infrastructure investment, customer incentive programs and services, procurement, and employment (Fig. 7.4). The metrics are also aligned with LADWP's priorities, which are tracked and reported to Los Angeles city officials, with more of the 50 identified metrics to be included as the EMDI program evolves and benefits from stakeholder input. LADWP stakeholders who are interested in better understanding the EMDI effort can find more information on LADWP's public website at LADWP.com/equitymetrics.

LADWP's Corporate Performance Group is responsible for gathering and managing the data for EMDI reporting. Data is regularly collected from program managers across the organization. The frequency of data collection ranges from monthly to semiannual, depending on the type of information. Much of the data is already being collected for other purposes. The program staff developed standardized reporting formats to better integrate EMDI reporting needs with existing reporting systems and streamline data management.

The data for each metric is aggregated by zip code and can be displayed geographically on one of three map layers showing boundaries by zip code, city council district, or neighborhood council area. The data itself can be presented as a heat map, as clusters or as numbers by zip code. A heat map layer showing poverty levels by census tracts can be turned on to assess implementation of programs relative to neighborhood economic demographics across the city. This layer is based on the

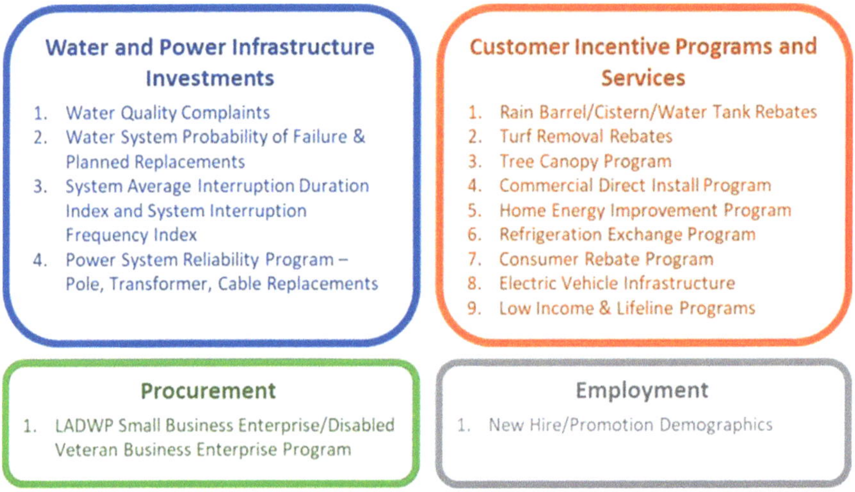

Fig. 7.4 Equity metrics

California Communities Environmental Health Screening Tool (CalEnviroScreen), a tool that the California Environmental Protection Agency (CalEPA) uses to designate California communities as disadvantaged [11]. Data tables are also provided for most metrics, providing additional detail.

EDMI reports are submitted to the Board semiannually and are made available to the public on the LADWP website. The report includes a map for each metric along with a dashboard. The dashboard summarizes the background, criteria, achievements, issues, and outreach strategy and plan for the program associated with each metric. The dashboard is an important tool to bring transparency and accountability to the operation units that are responsible for overseeing the various programs and services at LADWP. Figures 7.5 and 7.6 show an example EMDI map and dashboard report for the Refrigerator Exchange Program. Complete sets of reports can be found on LADWP's public website, LADWP.com/equitymetrics.

EMDI is an evolving process. As we go through the public outreach process, the public can review the semiannual reports, participate in stakeholder workshops, and provide input to the data initiative. The comments have been positive with respect to the development of the tool and the transparency that it brings to the public. Stakeholders have noted that these metrics allow both LADWP and community partners to strategically invest resources in areas with the most need. Recommendations for improvements are also being made. Stakeholders have asked that LADWP articulate clear goals for programs, show cumulative progress between reporting periods, and engage academic parties to conduct more detailed analysis of the data.

LADWP evaluates the findings and will continue to do so regularly. Program managers are able to use the data to assess the performance of their programs in comparison to stated goals and set additional goals where needed. They can make changes to programs and continue to follow results to determine the effectiveness of

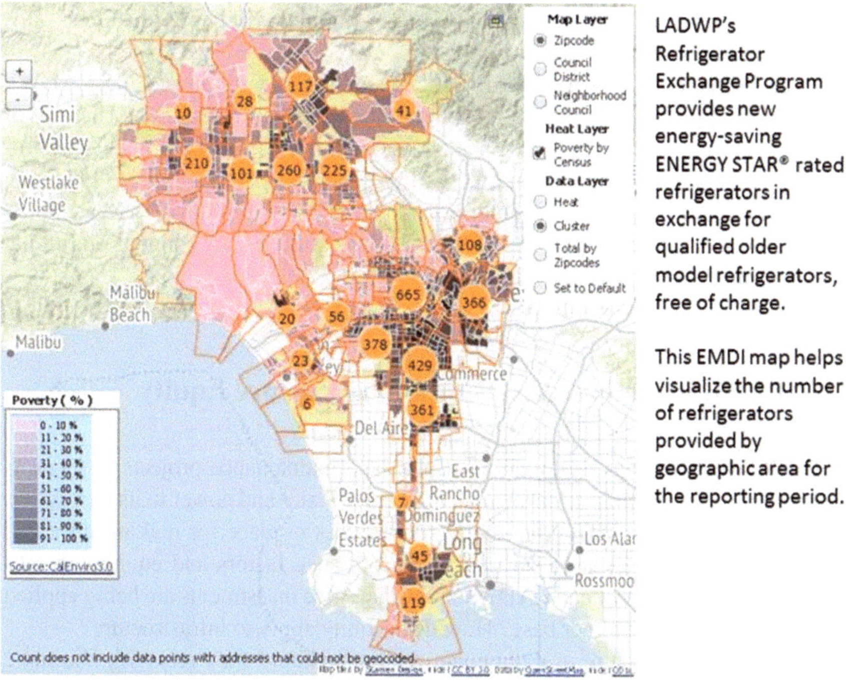

Fig. 7.5 EMDI – Refrigerator Exchange Program

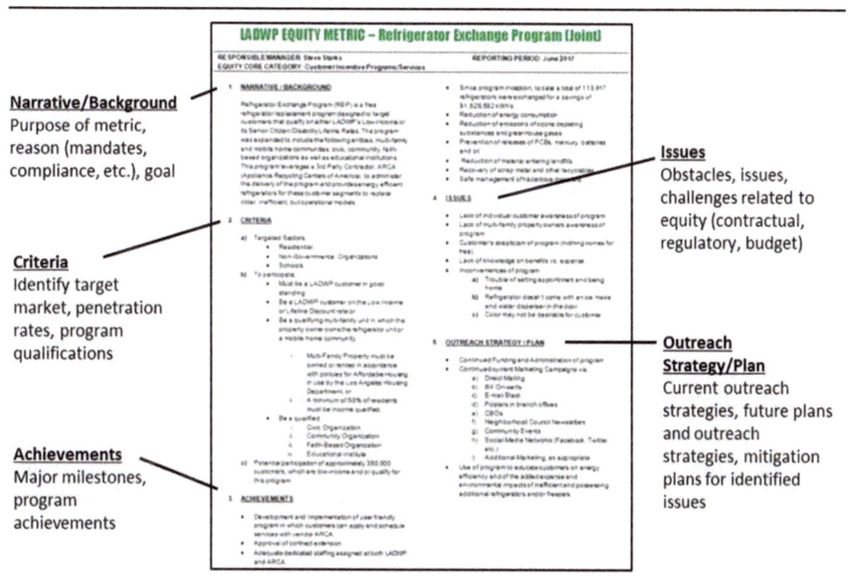

Fig. 7.6 Sample equity metric dashboard, Refrigerator Exchange Program

those changes. As the EMDI program proceeds, policies will be set to address variances from goals and performance gaps.

After introducing EMDI, Board Vice President Funderburk and the LADWP Office of Corporate Performance launched a broad outreach program. The aim of the outreach program is to educate stakeholders, gain input on EMDI, engage program managers, and further refine the EMDI program. LADWP is also engaging researchers and educators, including partnerships with Loyola Marymount University and the University of California, Los Angeles, to conduct additional analyses and provide feedback on the tool.

7.4 EMDI in Action: Assessing Infrastructure Equity

LADWP invests billions of dollars in capital and maintenance projects to meet its goals of providing reliable, high-quality, affordable water and power to its customers. As with all utilities, LADWP has standard practices in place for making informed infrastructure decisions based on system performance factors and equipment reliability indicators; yet how can LADWP show that these investments are being applied equitably across its customer base? How does equity apply to infrastructure?

Equity in infrastructure is about ensuring that customers are receiving the same level of service no matter what neighborhood they live in. It is about geography rather than socioeconomic conditions. Infrastructure equity often means a higher level of investment in older areas of the city, in areas with rapid growth, and in areas with special circumstances. For some areas of the city, high infrastructure needs do not track with low socioeconomic conditions.

From the system reliability perspective, LADWP considers many factors when making decisions on infrastructure investments. Some of these factors include equipment age, criticality to the system, equipment condition, frequency of maintenance, likelihood of failure, costs, and available resources.

From the customer perspective, we know from stakeholder meetings that the primary factors of interest are reliability, quality of service, and affordability. Equity is also important. Residents experiencing power interruptions or water quality issues want to know that their concerns are being heard and that their needs are being treated fairly. They want to know that whether they live in South LA, Westside, San Fernando Valley, Downtown, or East LA, investments in corrective actions and preventative maintenance are being made by their utility in a fair manner.

EMDI provides tools for evaluating infrastructure equity by tracking replacement of power poles, transformers, and cables; replacement of mainlines, trunk lines, and other water infrastructure; and water quality complaints. The System Average Interruption Duration Index (SAIDI) and System Average Interruption Frequency Index (SAIFI) are also tracked.

7.5 Visualizing Issues and Solutions

In early 2016, a number of complaints were made by customers in the South LA area who were experiencing intermittent cloudy water. As part of Equity Metrics, the Water Quality Division developed maps to assist with evaluating trends or identifying potential system issues. The map for FY 2015–2016 shows the location and intensity of the complaints. The problem was resolved in June of 2016 and is reflected in the map for FY 2016–2017, which shows a relatively low number of customer calls and no unusual patterns (Fig. 7.7). This information was known to LADWP management, but the value is that by publishing the maps through the EMDI program, management can visually show both the Board and the public that the issue was resolved in that time period by showing that customer complaints had dropped to normal levels.

7.5.1 Monitoring Power System Reliability Indicators

Power reliability indicators, System Average Interruption Frequency Index (SAIFI) and System Average Interruption Duration Index (SAIDI), are also being tracked through EMDI. SAIFI measures the average number of sustained interruptions per year experienced by the average customer, and SAIDI measures the average interruption time in minutes that the average customer experiences. These two indices are combined to accurately reflect the reliability performance of the power distribution system.

Prior to EMDI, these indices were measured with respect to relatively large geographic areas. To better understand customer impacts with EMDI, the data gather-

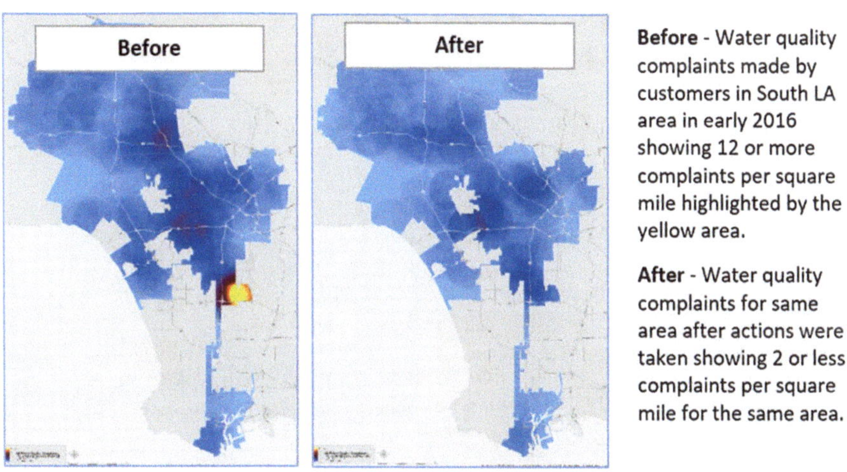

Fig. 7.7 Water quality complaints before and after action taken

ing methodology for SAIDI and SAIFI was changed to increase its geographic granularity. This resulted in a 1000-fold increase in the number of data points from 10,000 to 10,000,000.

LADWP's power system uses the SAIDI and SAIFI EMDI maps to assess the distribution of interruption duration and frequency values by geographic area and to assess progress in achieving targets for these key performance indicators. By doing so, LADWP is able to track geographic equity by understanding whether one neighborhood is experiencing an increased proportion of interruptions compared to other neighborhoods. This information is being used as a factor in prioritizing equipment replacement and as a means of measuring replacement program effectiveness. It also enables the customer to assess how their neighborhood is being served in the context of LADWP customer areas as a whole and with respect to LADWP's performance goals set out in its Power System Reliability Plan. The maps in Fig. 7.8 show the SAIDI numbers for various areas of the city.

7.6 Evaluating Customer Incentive Programs and Services

The changing nature of energy and water utility business models necessitates concurrent changes in the incentive programs and services. Sustainability initiatives, renewable energy, efficient products, watershed enhancement, and city sustainability goals are driving demand for new programs. Emerging technologies such as solar and electric vehicles are driving demand for access to distributed infrastructure. Increasing rates highlight the importance of helping customers lower their bills

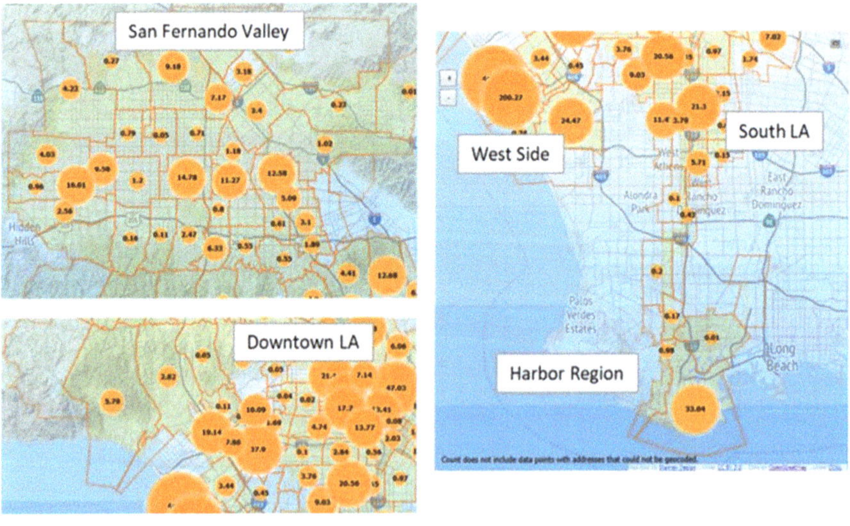

Fig. 7.8 System Average Interruption Duration Index

through efficiency measures and providing a safety net for low-income and senior customers (Fig. 7.9).

7.6.1 Reaching Diverse Customer Sectors

LADWP redesigned its portfolio of efficiency solutions programs in 2012 with the express purpose of making energy and water efficiency offerings available to all of its rate classes and market segments, including residential and nonresidential programs. It would be easy to design efficiency programs for the highest users, those with money to spend on solutions, and easy-to-reach customers; but that would leave out large segments of LADWP's customer base. EMDI is used by the efficiency solutions team to evaluate program performance and modify programs as needed to achieve the portfolio goals of equitable access to efficiency opportunities, promotion of green jobs for the local workforce, building capacity for community groups to promote efficiency, and operating transparently.

LADWP has a portfolio of efficiency programs meant to reach a diverse customer base, including technical assistance programs for large industrial and commercial customers, direct installation programs for smaller commercial customers, special outreach programs for hard-to-reach market segments, and a range of incentive programs with offerings for homeowners, renters, businesses, property owners, and institutions. While diverse program coverage is an important goal, programs must also meet cost-effectiveness standards in fairness to all ratepayers (Fig. 7.10).

David Jacot, Director of Efficiency Solutions, LADWP

"While LADWP has a goal to provide equitable energy efficiency programs to its customers, it also has a responsibility to ensure that its programs are cost effective. Programs that focus on specific sectors may sometimes fall below a cost-effectiveness ratio of 1.0. In order to fund these programs, LADWP balances cost effectiveness across the entire portfolio. This enables using above the line return-on-investment programs to fund below the line public benefit programs. In looking at the total resource cost for LADWP's energy efficiency portfolio, it is clear that even though some programs fall below the Total Resource Cost (TRC) of 1.0, the portfolio of energy efficiency programs exceeds the TRC target overall."

Fig. 7.9 Equity tradeoffs – balancing cost-effectiveness. (Personal communication, 1 December 2017.)

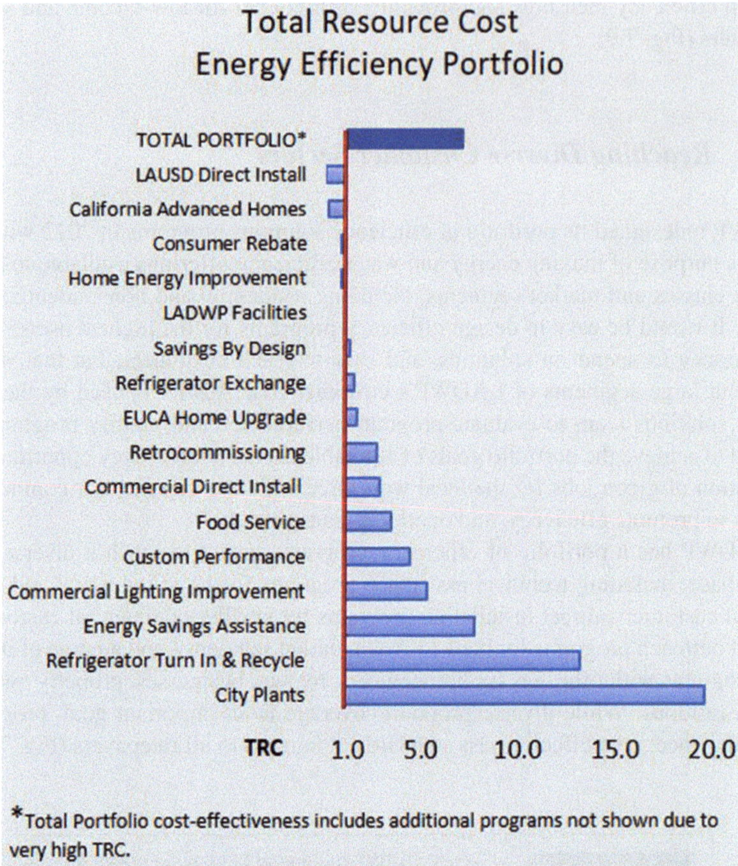

Fig. 7.10 Energy efficiency portfolio cost-effectiveness

One of the measures that LADWP uses to determine the cost-effectiveness of our programs is Total Resource Cost (TRC). The TRC is reported as a benefit-cost ratio that gives an indication of the rate of return of a program to the utility and its ratepayers. This test includes the effects of the program on both participating and non-participating customers and includes utility costs paid by both the utility and participants [12]. A TRC above one indicates that the program is beneficial to the utility and its ratepayers on a total resource cost basis.

The Consumer Rebate Program (CRP) and the Home Energy Improvement Program (HEIP) provide a good example of complementary programs that meet the needs of two different demographics. CRP is designed to reach out to residential customers to encourage them to purchase and install qualifying products in their homes. Outreach is general, and participation in the program is self-selected by the customer. HEIP provides free installation of selected energy and water efficiency measures for single-family and multifamily residential units. The strategy for this program is to prioritize low-income customers.

Comparing the heat maps for CRP and HEIP, where darker areas represent higher participation, shows how well these programs complement one another with respect to serving LADWP’s diverse customer base. As seen in Figs. 7.11 and 7.12, the heat map for CRP shows nearly the reverse distribution of participation compared to HEIP, highlighting the need for diverse programs to meet diverse customer capabilities. CRP participation is concentrated in areas with higher income levels, such as the San Fernando Valley in the north and the Westside, while HEIP offsets this distribution by being heavily concentrated in the lower-income areas of South LA, East LA, and the northeast San Fernando Valley. The HEIP distribution also indicates that the strategy of prioritizing low-income customers is succeeding, as these geographic locations are also identified as economically disadvantaged per CalScreen.

In addition to CRP and HEIP, five other efficiency programs are tracked through EMDI. These include rebates for rainwater collection barrels, cisterns, and tanks,

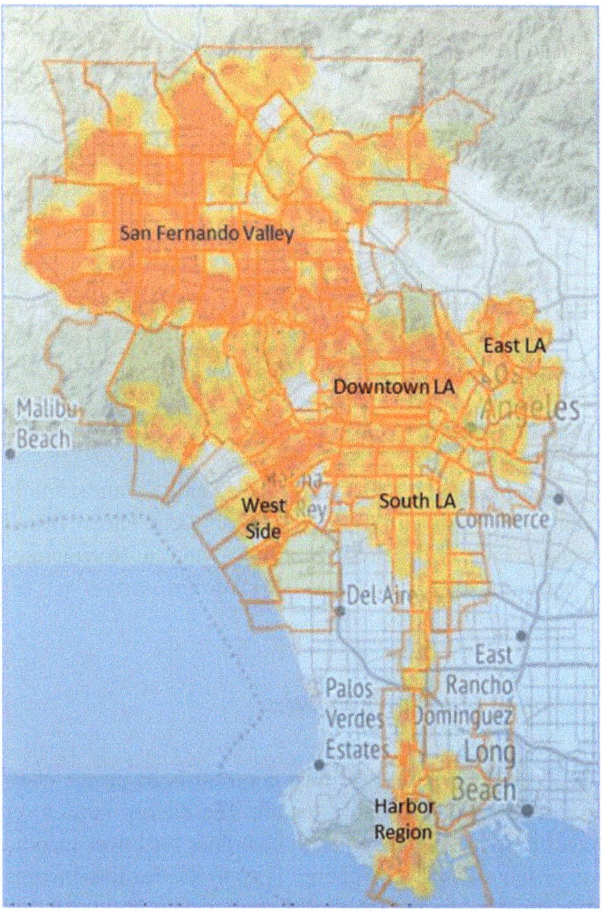

The EMDI map for CRP shows a high concentration of participation by customers in the moderate to upper income areas of the San Fernando Valley, northern Downtown LA, and the West Side. This program offers incentives to residential customers who implement energy efficiency solutions for the home.

Fig. 7.11 Consumer Rebate Program participation

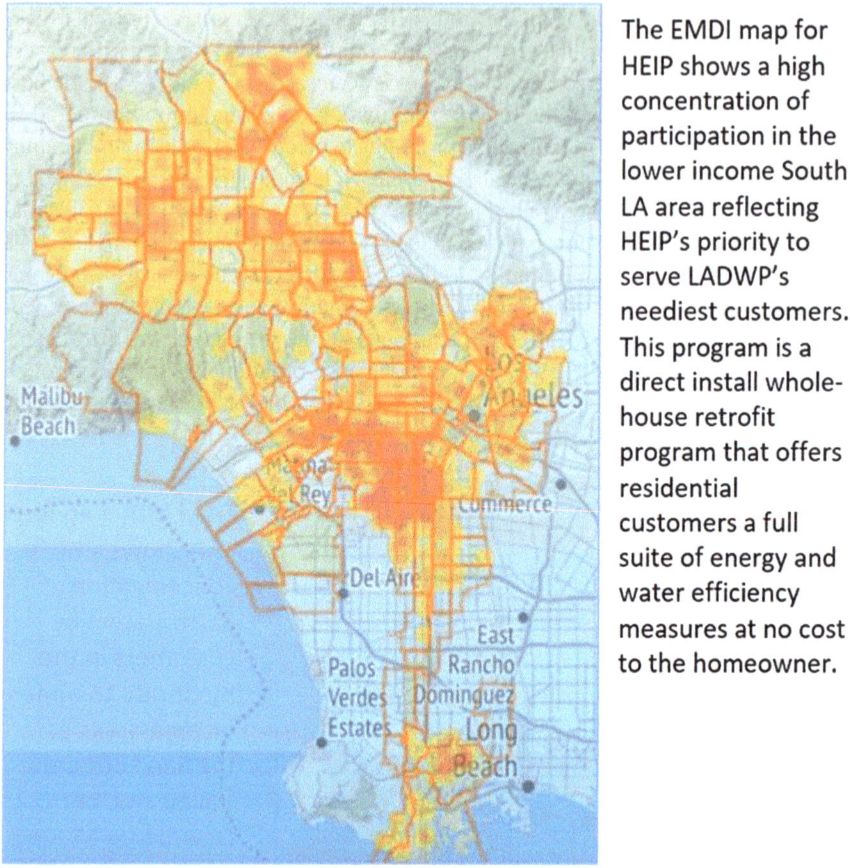

Fig. 7.12 Home Energy Improvement Program participation

incentives for turf removal and replacement with California Friendly® landscaping, a free tree program to increase tree canopy in the city, a Commercial Direct Install Program for small businesses and multifamily buildings, and a Refrigerator Exchange Program. Other programs may be added as EMDI progresses.

7.6.2 Measuring Customer Outreach Results

As LADWP advances a clean energy vision, we must confront head-on the challenge of ensuring that clean energy is affordable for all. The energy burden, or proportion of household income spent on energy, is much higher for lower-income households. The proportion of income spent on energy is up to 8% for low-income families compared to 3% for middle-income families. Families with higher energy burdens also experience many negative long-term effects on their health and well-being [13]. LADWP offers two discounted rate programs to make water and electricity more affordable to qualifying families who are experiencing difficulties paying their bills, the Lifeline Discount Program (LDP) and the Low-Income Discount Program (LIDP). LADWP has partnered with the City of Los Angeles Office of Finance to offer customers who are 62 years of age or older or permanently disabled a discount on their electric and/or water bills through the LDP (Fig. 7.13). LIDP offers a low-income discount rate for customers with qualifying income levels (Fig. 7.14). For both programs, renters pay for electric service only and are eligible to receive the low-income water subsidy credit.

The primary objective of these programs is to reach as many people as possible who qualify. One of the barriers to participation is people not knowing about the programs and not knowing that they are eligible. Another is a fear of registering for the programs, especially among non-English-speaking customers. In an effort to address these barriers, LADWP is planning to increase partnering with community-based organizations to conduct outreach and register customers. Measuring results for these types of projects has always been a challenge. EMDI is an additional tool

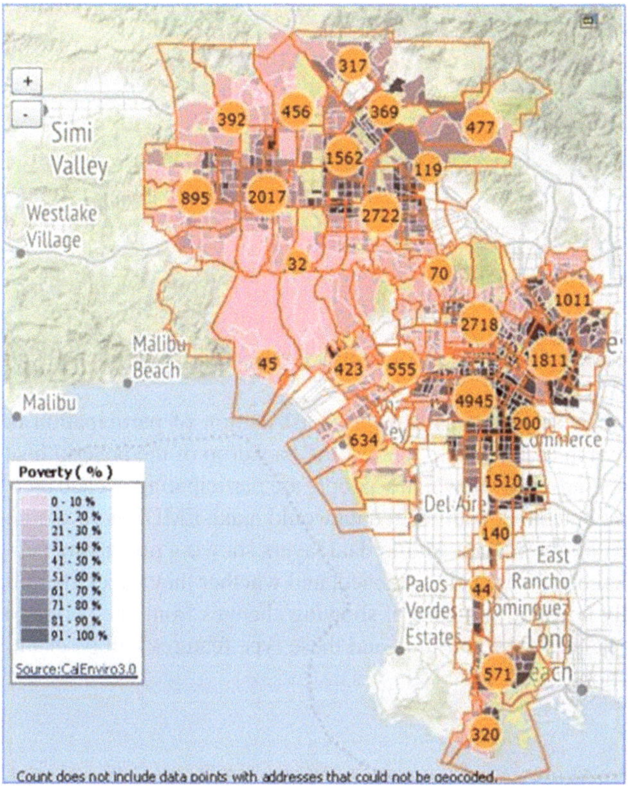

This map provides a snapshot of the number of households participating in LDP in the month of June 2017. This program offers customers who are 62 years of age or older or permanently disabled a discount on their electric and/or water bills.

Fig. 7.13 Lifeline Discount Program

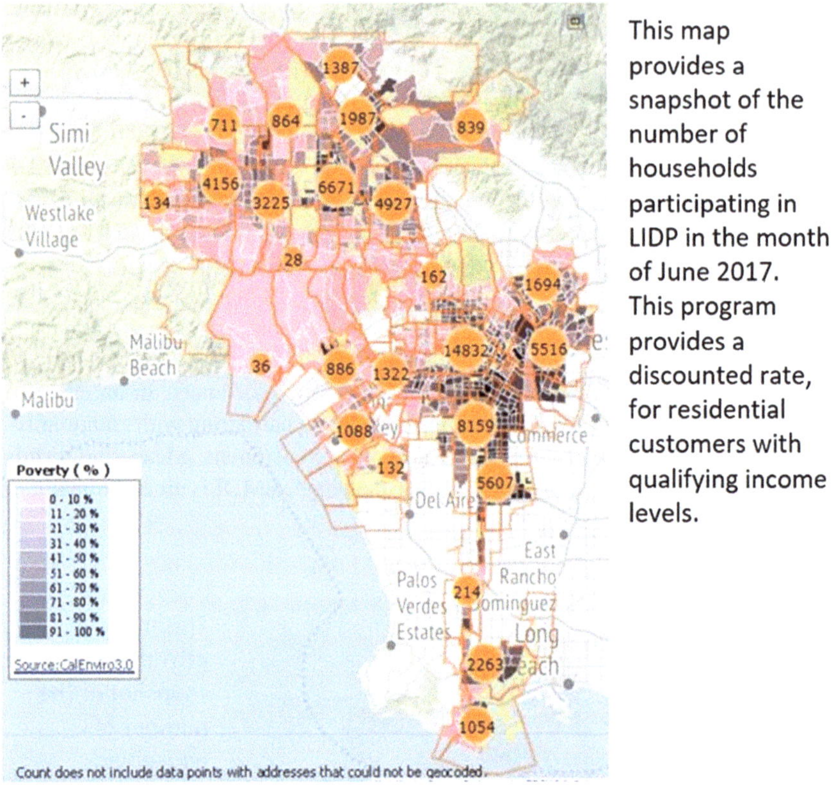

Fig. 7.14 Low-Income Discount Program

that can be used to assess the effectiveness of contractors and partner groups within their target areas.

The EMDI maps for these programs show the distribution of participation in each program for the month of June 2017 against the backdrop of the poverty heat layer. As it stands, these maps show how many people are participating in each area. Stakeholder groups have noted that additional data would make EMDI more useful. For these programs, this might include adding data layers showing how many people are eligible, what percentage are being reached, and whether they reside in multifamily or single-family housing. A means of showing changes from one period to the next would also be helpful. The ability to add these type features will be examined in future modifications.

7.6.3 Investing in Disadvantaged Communities

"Until every car, bus and truck has a plug and every plug is powered by the earth, the wind and the sun, families will continue to feel the perils of pollution. While many wealthy parts of LA have an abundance of electric vehicle chargers, we're not seeing the same investments in communities like Watts. That needs to change."[2] This statement by Reverend Mac Shorty of the Watts Neighborhood Council in South Los Angeles highlights the importance of an equitable approach to electrification. Electric vehicle (EV) programs need to make EV ownership and drivership accessible to residents throughout the city.

The initial rollout of personal electric vehicles required the purchase of a new vehicle and installation of a home charger, both expensive investments for the average consumer. The need for a home charger also made this type of purchase out of reach for renters. Meanwhile, with previously owned vehicles hitting the used car market and manufacturers offering EV options in lower-cost models, the cost of EV ownership is becoming more affordable for the average consumer.

As LADWP expands its investment in EV infrastructure and enters into new programs with other entities to expand EV accessibility in disadvantaged neighborhoods, EMDI will be used to track the geographic location of installations. This information can be used to evaluate investment equity across the city and to modify or target additional programs to address inequities.

In 2016, LADWP installed a utility pole-mounted EV charger next to a community service center in the Watts neighborhood as part of a pilot program to expand EV infrastructure citywide (Fig. 7.15). This was a significant advancement in that

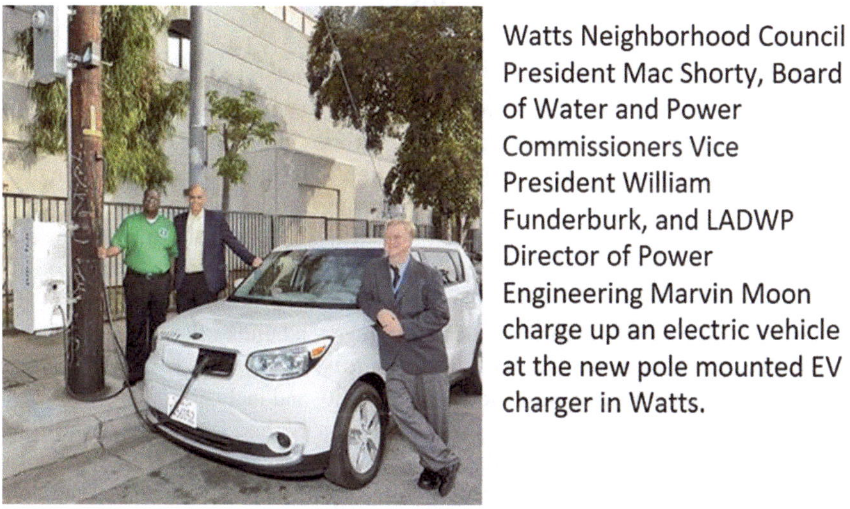

Fig. 7.15 Pole-mounted charger (photo August 2017)

[2] Personal communication, 1 December 2017.

LADWP-installed power pole chargers are simple to install using the existing wires on the power poles and avoid breaking up streets or sidewalks for underground wiring. Additional pole-mounted chargers are planned for other locations as part of this program.

The City of Los Angeles recently authorized a unique electric vehicle car-sharing pilot project, called Blue LA, that focuses on disadvantaged neighborhoods in Central Los Angeles. The city is leading the program with a grant from the California Air Resources Board, and LADWP is providing funding in the form of rebates for the charger infrastructure. Up to 200 curbside chargers will be installed in disadvantaged communities as part of this project.

Using a grant from the South Coast Air Quality Management District, LADWP is also partnering with Clean Fuel Connection and NRG EVGO on another project to install a managed EV charger network in the city. Clean Fuel Connection is installing the chargers on city property. NRG EVGO will operate the charger system, collect revenues, and pay the monthly electrical bill to LADWP. This approach streamlines ownership and management of a network of chargers, saving resources on the part of LADWP and incentivizing private investment in the charger infrastructure.

Moving forward, LADWP will use the poverty heat map of EMDI to target, track, and ensure the equitable distribution of EV infrastructure throughout LADWP territory. This information – as with all other programs – will not just inform program managers but will also be reported to LADWP's Board of Commissioners and shared with LADWP customers online.

7.7 Valuing Diversity and Inclusion

LADWP is a major employer and purchaser of goods and services in the Los Angeles region (Fig. 7.16). Many prospective employees and vendors compete with each other to get an LADWP contract. Inherent biases in cognitive decision-making can result in human capital and corporate vendors being either overvalued or undervalued. LADWP tries to even things out by having policies and initiatives to purchase goods, services, and labor equitably. This encourages competition and diversity, which increases corporate efficiency and effectiveness. According to Bill Funderburk, Vice President of the Board of Water and Power Commissioners, "In essence, equity metrics is a form of social arbitrage – it builds value in the organization by leveraging the synergies of diverse resources. EMDI helps LADWP evaluate its progress while emphasizing its focus on diversity and inclusion."[3]

LADWP is expanding its existing data collection process for contracts and contractors to include more granular data to provide information about the equity of contract allocation according to several metrics. These include the number and dollar value of contracts awarded to small business (SBE), women-owned (WBE),

[3] Personal communication, 1 December 2017.

Fig. 7.16 LADWP vendor fair

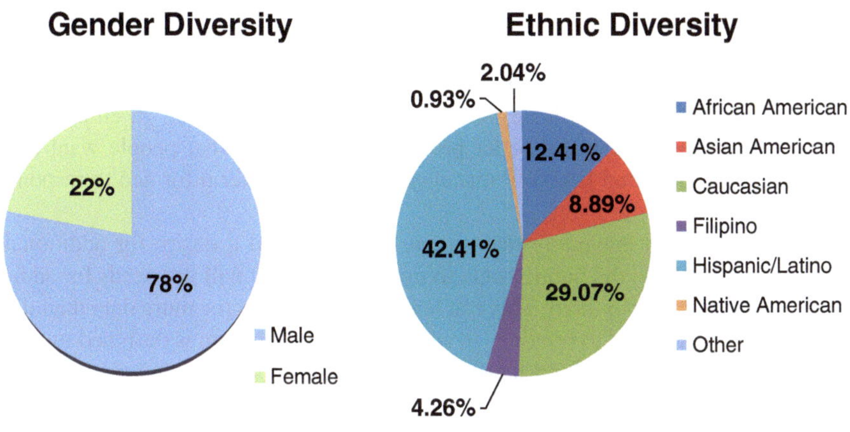

Fig. 7.17 LADWP employment demographics

minority-owned (MBE), disabled veteran-owned (DVBE), and LGBT-owned business enterprises; business locations; and industry category. EMDI will be expanded accordingly to track these parameters and identify areas needing improvement. Current vendor data shows LADWP is exceeding its SBE goal of 25% overall in contracts eligible for SBE goals (Fig. 7.17). The EMDI dashboard lays out an outreach and strategy plan calling for an informal increase in this goal to 30% and outlines the outreach activities planned for FY 2017–2018.

For employment metrics, LADWP is expanding its existing data collection framework to include information to evaluate the equity of training and hiring practices according to the following metrics: gender, ethnic background, disabled veteran status, date of hire, residential location, and educational level. In addition, new targeted recruitment initiatives have been instituted to increase female and minority candidates for engineering and critical craft job classifications. These targeted initiatives include attending gender/diversity organization national conferences, expanding college recruitment to additional campuses, consulting with other city departments and utilities, and expanding military outreach and recruitment. Going forward, EMDI will be used to measure progress in these areas.

7.8 Tradeoffs

Overall EMDI is a beneficial program; but as with anything, there are tradeoffs.

Sharing equity data is powerful. It impacts how customers see LADWP and how LADWP interacts with customers (Fig. 7.18). While EMDI gives LADWP the ability to present data in context using a tool that is clear and understandable and that can be used to objectively evaluate the equitable distribution of resources, it also gives the public leverage to hold the Board and the organization accountable.

Transparency invites public feedback – both positive and negative. Stakeholders are able to evaluate the distribution of resources and program participation in their areas. They may use the data to call out disparities and to make recommendations for improvements. They will also be able to see the success or lack of success for program changes that have been made in response to their input. In either case, public feedback will result in better programs that reflect what people want and need, and the Board and LADWP management will be better informed to respond to the public.

New information inevitably raises more questions and a desire for additional data. As LADWP and the public begin to use the data, they will see needs for additional data that can help them. This can lead to expectations for more data than the organization is prepared to provide. To address this, the program is designed to give the user access to the data and leave it up to the user to analyze and interpret the

John Chen, Director of Corporate Performance, LADWP

"Achieving the best results with Equity Metrics is an art, not an exact science. There's a constant need to strike a balance between what we dream of having versus what we can deliver. It is a continuous iterative process in open engagement with our stakeholders throughout the City to ensure that our programs and services are meeting their expectations. Equity is also about equity in opportunity, allowing access for everyone so that we can build a diverse workforce and suppliers."

Fig. 7.18 Equity metrics – it's an art. (Personal communication, 1 December 2017.)

data. LADWP employees will be given access to detailed data and will be able to conduct analyses of interest to them. LADWP will also be partnering with other entities to give them access to detailed data for further analysis. In providing access to data for researchers as well as the general public, it will be important to strike a balance between transparency and data privacy and security.

The program opens the door to a more expansive project requiring additional work and resources. One aspect of EMDI is to increase the granularity of data collection. For one metric, this meant going from collecting 10,000 data points to collecting 10,000,000. While this requires extra staff resources to set up, the additional information is also useful to LADWP operations and maintenance. Future improvements to the program and fulfilling requests for additional data will be governed by resource availability.

No matter the tradeoffs, it is always better to operate from transparency. LADWP wants to provide meaningful and impactful programs and services to its customers. This requires consideration of many factors when prioritizing how to invest limited resources. Measuring and monitoring equity metrics is a tool to do so.

7.9 Conclusion

According to the City Sustainability Plan, by 2035, at least 500,000 more people will be calling LA home. The majority of these residents, the millennials, are defining the shared economy and the Internet of Things. As much as they value connectedness, they place an emphasis on their independence. They are likely to have solar panels to take care of their electricity needs, ride share or use autonomous electric vehicles, and are bound to make decisions toward energy independence.

We posit that LADWP is changing with the times to ensure that it continues to grow with and share the same values as its customers. As LADWP moves forward, EMDI is an essential, replicable, and scalable tool to make sure that we are serving our customers fairly (Fig. 7.19). It enables visually tracking results and focusing

Nancy Sutley, Chief Sustainability Officer, LADWP

"As organizations struggle to figure out how to promote equity, LADWP has developed a meaningful tool. Setting up your own equity metrics system is not hard to do. The first step is to decide to do it. You need to want to know the answer. Pick enough data so it is meaningful and don't be afraid to see the results. The information will prove to be useful in many ways. Use data that is readily available and find a way to display it. Ask yourself what issues are affecting people in your service area; then make the data talk to these issues."

Fig. 7.19 Equity metrics – a tool for today's utility. (Personal communication, 1 December 2017.)

efforts to achieve equitable outcomes that balance the needs of the utility, our stakeholders, and our customers. The challenge for LADWP now is to achieve these city-wide policy aims while still meeting the primary goals of the organization – providing reliable, safe, clean, and affordable water and power with excellent customer service while flexing with the larger transformation that is happening in the utility industry.

Acknowledgements The authors would like to thank LADWP's staff, management, and counsel for the work they put in to developing the equity metrics. LADWP communications staff were supported by Susan Campos and Brad Kane. The authors would also like to thank Maria Sharma, Aditya Sharma, Carolyn Casavan, and the LADWP communications team for their support in preparing this chapter. The website for equity metrics is www.ladwp.com/equitymetrics.

Author Profiles

Nancy Sutley Sutley is LADWP's Chief Sustainability Officer. Sutley oversees LADWP's energy efficiency, water conservation, environmental affairs, electrification of the transportation network, and the LaKretz Innovation Campus. Sutley served as Chair of the White House Council on Environmental Quality from 2009 to 2014. Under Sutley, the Council on Environmental Quality played a central role in shepherding the Obama Administration's signature environmental accomplishments. Sutley was one of the chief architects of President Obama's June 2013 Climate Action Plan to reduce carbon pollution, prepare the United States for the impacts of climate change, and lead international efforts to address climate change. Prior to her appointment as Chair of the Council on Environmental Quality, Sutley was the Deputy Mayor for Energy and Environment for the City of Los Angeles, California, under Mayor Antonio Villaraigosa. She represented Los Angeles on the Board of Directors for the Metropolitan Water District of Southern California. She served on the California State Water Resources Control Board from 2003–2005. Sutley also was Energy Advisor for California Governor Gray Davis and Deputy Secretary for policy and intergovernmental relations at the California EPA from 1999 to 2003. During the administration of President Clinton, Sutley worked for the US Environmental Protection Agency as a Senior Policy Advisor to the Regional Administrator in San Francisco and Special Assistant to the Administrator in Washington, D.C. Sutley received her bachelor's degree from Cornell University and her Master of Public Policy degree from Harvard University.

William Funderburk Funderburk is an environmental lawyer with a keen understanding of how to deal with the critical issues that face utilities today. A key figure in Los Angeles, Funderburk has served as the Vice President of the Board of Water and Power Commissioners at LADWP since 2013. He is managing partner of Castellon & Funderburk LLP in Los Angeles, which he co-founded in 1999. Funderburk is known for being a visionary when it comes to sustainability, renewables, clean tech, water quality, and equity. Funderburk introduced the Equity Metrics Data Initiative at LADWP and continues to play a key role in its implementation. He has published various articles regarding California stormwater regulations, clean power and distributed generation, Superfund reform, and insurance recovery for environmental liabilities. He is also a much sought-after speaker on environmental compliance. Funderburk is a member of the California State Bar and District of Columbia Bar. He graduated from Yale University with a Bachelor of Arts degree in Engineering and Applied Science and from Georgetown University Law Center with a JD degree. He was born in Washington, D.C., and lives in Los Angeles, California, with his wife and two daughters.

John Chen Chen has worked for LADWP for over 27 years. He started working as a student engineer and held various technical and senior management positions at the company overseeing rates and retail contracts, telecommunication networks, marketing and customer services, and economic development. He is currently the Director of LADWP's Corporate Performance Office.

Chen served as a member of Mayor's Economic Development Cabinet, a member of the Citywide Los Angeles River Revitalization Task Force, and a member of Board of Governors for the Los Angeles Economic Development Corporation. He is currently the acting President of the Management Employee Association at LADWP. He was one of the founding members that purchased the property on behalf of LADWP in creating the LA Cleantech Center. Chen led the consortium of academic institutions from USC, UCLA, and Caltech, along with LADWP, that applied and won the $120 million smart grid grant from the Department of Energy. He has a bachelor's degree in Electrical Engineering, a master's degree in Power Engineering, and an MBA from the University of Southern California.

David Jacot, P.E. Jacot oversees all aspects of LADWP's offerings and strategies designed to overcome market barriers to the comprehensive adoption of energy efficiency by LADWP's customers, as well as the implementation of LADWP's class-leading water conservation and efficiency programs. Jacot also oversees the integration of water and energy efficiency program delivery across LADWP's service territory as well as through a nation-leading joint program partnership with the natural gas utility serving Los Angeles, the Southern California Gas Company. Jacot has a bachelor's degree in Mechanical Engineering from the University of Oklahoma and a master's degree in Urban and Regional Planning from California State Polytechnic University, Pomona, as well as 15 years of experience in designing high-performance building systems, modeling building energy usage, and managing cost-effective and investment-grade energy efficiency programs.

Maria Sison-Roces Sison-Roces serves as a Utility Services Manager for LADWP's Office of Sustainability. Sustainability, integrated reporting, ordinance compliance, and cross-disciplinary engagement are the focus of her work experiences in the last 5 years. Sison-Roces manages LADWP's compliance with the existing building energy and water efficiency ordinance and environmentally preferable purchasing ordinance. She is also responsible for developing and implementing green initiatives to elevate employee awareness on sustainability. In 2016, Sison-Roces led cross-functional efforts to achieve LEED Gold Certification for LADWP's headquarters in Downtown Los Angeles. She has a Bachelor of Arts degree in European languages from the University of the Philippines, a master's degree in Business Administration from the University of La Verne, and a sustainability certificate from the University of California, Los Angeles. Sison-Roces also serves as co-chair for the Publicly Owned Utilities Working Group and P198 Strategic Sustainability Science for the Electric Power Research Institute's Energy Sustainability Interest Group.

Company Profile

Los Angeles Department of Water and Power (LADWP) is the largest municipal utility in the United States and the third largest utility in California. LADWP serves 680,000 water customers and 1.5 million electric customers. LADWP has over 7,880 MW of generation capacity from a diverse mix of energy sources including renewable energy (29%), natural gas (34%), nuclear (9%), large hydroelectric (3%), coal (19%), and unspecified sources (6%). The first California utility to reach 20% renewable energy in 2010, LADWP is on track to meet the state-mandated requirement of 33% renewables by 2020. LADWP is committed to obtaining 15% of projected power needs in 2020 from energy efficiency, which means that almost half of the electric supply in Los Angeles in 2020 will come from renewable resources. Increasing renewable energy, replacement of coal, infrastructure reliability investments, ramping up energy efficiency, and other demand-side programs are critical and concurrent strategies being undertaken by LADWP. In 2016, LADWP's revenues were $4.7 billion, and they employed approximately 9,700 employees.

References

1. "Census Bureau Reports at Least 350 Languages Spoken in U.S. Homes," United States Census Bureau Press Release, https://www.census.gov/newsroom/press-releases/2015/cb15-185.html, Release Number: CB15-185, 3 3 November 2015. Accessed 25 January 2018.
2. Galperin, Ron, "Owner-Occupied Housing, 2015," EconomyPanel LA, Los Angeles City Controller, http://www.lacontroller.org/economypanel_la. Accessed 25 January 2018.
3. "Quick Facts, Los Angeles City, California," United States Census Bureau, https://www.census.gov/quickfacts/fact/table/losangelescitycalifornia/PST045216. Accessed 25 January 2018.
4. "Census Bureau Reports at Least 350 Languages Spoken in U.S. Homes," United States Census Bureau Press Release, https://www.census.gov/newsroom/press-releases/2015/cb15-185.html, Release Number: CB15-185, 3 November 2015. Accessed 25 January 2018.
5. "Quick Facts, Los Angeles City, California," United States Census Bureau website, https://www.census.gov/quickfacts/fact/table/losangelescitycalifornia/PST045216. Accessed 25 January 2018.
6. "Less Than High School," Mapping L. A., http://maps.latimes.com/neighborhoods/education/less-than-high-school/neighborhood/list/, Los Angeles Times, June 2010. Accessed 25 January 2018.
7. "LADWP Equity Metrics Data Initiative Report," www.ladwp.com/equitymetrics, Los Angeles Department of Water and Power, 1 August 2017, p. 22.
8. "From the Street: Civic Epistemologies of Urban Climate Change," by M. Mendez in *Spatializing Politics: Essays on Power and Place*, Harvard University Press (http://www.hup.harvard.edu/catalog.php?isbn=9781934510469)

9. "Assessing local climate action plans for public health co-benefits in environmental justice communities," by M. Mendez in *Local Environment: The International Journal of Justice and Sustainability* (https://www.tandfonline.com/loi/cloe20)
10. "Additional Proposed Equity Metrics for Considerations (Executive Summary)," www.ladwp.com/equitymetrics, Los Angeles Department of Water and Power, December 2016.
11. "CalEnviroScreen," https://oehha.ca.gov/calenviroscreen, California Office of Environmental Health Hazard Assessment (OEHHA). Accessed 25 January 2018.
12. "California Standard Practice Manual: Economic Analysis of Demand-Side Programs and Projects," California Public Utilities Commission, October 2001. p. 18.
13. Drehobl, Ariel and Ross, Lauren. "Lifting the High Energy Burden in America's Largest Cities: How Energy Efficiency Can Improve Low Income and Underserved Communities," American Council for an Energy-Efficient Economy, April 2016. pp. 9, 13.

Chapter 8
Portland General Electric Driving Collaboration for Sustainable Legislation

Dave Robertson and Brendan McCarthy

Abstract In 2015, a group of environmental organizations filed a ballot measure petition that would require investor-owned electric utilities in Oregon to acquire 50% of their energy from renewable resources by 2040 and cease serving customers with coal-generated electricity. To mitigate the potential economic and social impacts of this measure, the two major investor-owned electric utilities in the state, Portland General Electric and PacifiCorp, formed a workgroup with utility stakeholders, customer advocates, renewable energy developers, and environmentalists. Together, they negotiated a new plan, Senate Bill 1547, which was adopted by the state legislature in February 2016. Compared with the measure, the bill significantly reduced the cost impacts of emission reductions, improved flexibility in meeting carbon reduction goals, created a community solar program, and allowed utilities to invest in transportation electrification using customer funds. This collaborative effort limited the total possible trade-offs involved in reducing carbon emissions while controlling customer costs.

8.1 Introduction

Portland General Electric Company (PGE) is an investor-owned utility located entirely within Oregon. Headquartered in Portland, PGE has provided electricity for more than 125 years and as of 2017 serves 872,000 customers in a six-county region in the Northern Willamette Valley. PGE wholly and jointly owns hydroelectric, natural gas, coal, and wind generation facilities that totaled 3843 MW of nameplate capacity in 2017, at a rate lower than the national average residential price per kilowatt-hour (kWh).

D. Robertson · B. McCarthy (✉)
Portland General Electric, Portland, OR, USA
e-mail: Brendan.McCarthy@pgn.com

J. Fox, M. Scott (eds.), *Sustainable Electricity II*,
https://doi.org/10.1007/978-3-319-95696-1_8

Oregon is home to several well-organized environmental groups with thousands of active members[1] [1, 2] and draws new residents from across the country who want to live and work in a beautiful natural landscape—and many bring a strong environmental stewardship ethic[2] [3, 4]. As such, PGE has the highest participation in the country (both by percentage and total number of customers) in its voluntary renewable power program. This program is open to both residential and commercial customers and has more than 150,000 participants, approximately 20% of PGE customers. In addition, customers consistently rank PGE in the top quartile of US electric utilities for protecting the environment [5]. At the same time, PGE serves both large and small customers who are concerned about their energy prices—including those in energy-intensive industries, such as semiconductor and computer chip manufacturing, wood products, and food processing—as well as significant numbers of low-income residential customers.

PGE has a history of collaborative engagement when it comes to finding solutions to politically charged issues and consistently strives to play an active role in addressing environmental concerns, especially those that involve energy development. Further, the company seeks not only to protect the environment but also the communities in which it operates and the customers who pay for environmental stewardship efforts through fees on their electric bills. PGE approaches sustainability through a foundation of "people, planet, and performance," in which sustainability is a driver for resource allocation, future investments, and procurement practices. These three focal points mirror the traditional "pillars" of sustainability in popular business theory: society, environment, and economy. Beginning in 2016, PGE's process for approval of all capital investments required the evaluation of sustainability issues alongside traditional business considerations, such as customer value, environmental impact, and financial benefit.

In 2016, PGE's efforts in sustainable planning were rewarded when the company played a key role in successfully negotiating legislation that would dramatically impact the electricity sector. As part of a workgroup, PGE helped draft the language of Senate Bill 1547 (SB 1547) to meet a critical environmental goal—the expansion of the Oregon Renewable Portfolio Standard (RPS) from a 25 to a 50% requirement—while mitigating the economic impact on both utilities and customers. The Oregon Legislative Assembly adopted SB 1547 in a short, 1-month legislative session once the language had been carefully drafted to include input from numerous parties. While the adoption of this bill took only a month, the groundwork had been

---

[1] The Oregon chapter of the Sierra Club claims 20,000 members and Renew Oregon claims that more than 700 businesses support taking action on clean energy.

[2] "'This year's data reflects longer-term trends of people moving to the Pacific West, where cities such as Portland and Seattle are seeing the combination of a boom in the technology and creative marketing industry, ***as well as a growing "want" for outdoor activity and green space***,' Michael Stoll, economist, professor and chair of the Department of Public Policy at the University of California, Los Angeles, said in the United Van Lines release" (emphasis ours). Furthermore, according to the Oregon Values and Beliefs Project, 2013, Oregonians say environmental awareness is the no. 1 reason Oregon will be a better place to live in 10 years, 74% believe water and air quality are important, and six in ten believe environmental protection should be given more priority, even if it slows economic growth.

laid months and years before, demonstrating that it is possible for the environmental community and the electricity sector to work together to accomplish durable, innovative, and sustainable policy objectives.

This chapter provides the perspective of a negotiation process from only one party. Multiple parties were responsible for the successful legislation that significantly reduces Oregon's climate change-causing emissions from the electricity sector. Consistent with the themes expressed in this chapter, this outcome was a group effort.

8.2 History of Emissions Reduction at PGE

In 2006, PGE's then-CEO Peggy Fowler publicly stated that it was time for a national, market-based mechanism to reduce carbon emissions from the electricity sector. Since then, PGE has consistently taken actions to support the reduction of climate change-causing emissions. In 2007, PGE was instrumental in the drafting and adoption of the Oregon RPS, which required PGE and other utilities to have 25% of their electricity, on average, come from qualifying renewable resources. In addition, that same year PGE supported legislation that increased the amount of energy-efficiency spending for residential and commercial customers. In 2009, PGE helped support the adoption of a solar photovoltaic (PV) capacity standard, which required the installation of utility-scale PV cells in the state and supported a bill to create a greenhouse gas emissions performance standard for thermal power plants. In addition, PGE supported attempts to adopt a national cap and trade program on the federal level.

8.2.1 Negotiating the Boardman 2020 Plan

In 2010, PGE reached a comprehensive agreement with the Oregon Public Utility Commission (OPUC), the Oregon Department of Environmental Quality (DEQ), and the federal Environmental Protection Agency (EPA) to cease burning coal at its Boardman Generating Station (Fig. 8.1) by the end of 2020. The coal facility is a 585 MW-generating asset that delivers approximately 12% of PGE customers' energy needs. At full power, it could deliver energy to more than 600,000 homes. The plant employs 110 full-time workers, contributes significantly to the tax base of a rural community, and has been a reliable workhorse in PGE's generating fleet since 1980.

The agreement reached with regulators, known as the Boardman 2020 plan, was made within the context of the Best Available Retrofit Technology (BART) review under the Regional Haze Rule of the Clean Air Act. Because of its construction date, the Boardman plant was grandfathered into the 1977 amendments of the Clean Air Act and thus was built without certain pollution controls to lower nitrous oxides and sulfur oxides. The Regional Haze Rule is a federal set of standards to lower visibility-impairing, haze-causing pollutants from electricity-generating sources. The BART process was designed to mitigate the emissions of grandfathered generating plants

Fig. 8.1 Boardman plant

such as Boardman. To comply, PGE would likely have had to install selective catalytic reduction devices for nitrous oxides and "scrubbers" for sulfur oxides. These devices, along with filters for mercury pollution, would have required investments of $520 to $560 million. To recover these funds from customers at a reasonable rate, the plant would have had to run for many additional years.

Instead, PGE proposed installing less expensive pollution control equipment, including low-nitrous oxide burners and dry sorbent injection for sulfur oxides, at a cost of less than $100 million, with an agreement to cease burning coal by the end of 2020. PGE showed that the overall contribution of pollutants to the atmosphere under the Boardman 2020 plan would be less than that of the long-term mitigation plan. In essence, this new strategy would use more modest controls and plant closure as a control technology. Both the EPA and state regulators agreed with this approach.

The Boardman 2020 plan proved the value of acting with a sustainable mindset. Had PGE not installed pollution controls at all, the plant would have had to close by January 1, 2014. This would have had drastic effects on the rural community of Boardman due to a sudden loss of tax base and employment, as well as potentially unacceptable replacement power costs. On the other hand, installing a full suite of controls presented significant financial and environmental risk, as the Boardman plant may have been operating as late as 2040. During that 30-year period, carbon costs would likely have been imposed. The agreement represented the best path forward for both PGE and its customers, compared with the installation of pollution controls and a potential carbon cost risk over the time of recovering the sunk investment.

A critical aspect of the Boardman 2020 plan was that in reaching the agreement, PGE forged close ties with a core group of local environmental and customer advocates. This demonstrated that PGE could be trusted to bring critical and innovative thinking to resolve difficult problems, especially those concerning the environment.

8.2.2 *Joining an Industry Movement*

In 2015, PGE signed the White House's American Business Act on Climate Pledge, joining companies such as Alcoa, Apple, Bank of America, Berkshire Hathaway Energy, Google, IKEA, Intel, Microsoft, Nike, and Target. The pledge supports strong regulation of climate change-causing emissions with specific, quantifiable steps to reduce emissions, increase low-carbon investments, and build and use more clean energy resources. In 2017, PGE signed the #wearestillin pledge, ensuring the company is still committed to acting on climate change in the face of national stasis on the topic.

All these actions cemented bonds between utilities and the environmental community in Oregon. While the relationship is not without conflict—due in part to inherent tension between environmental regulation and business operations—the constant dialogue and engagement during and between policy efforts allow all parties to find opportunities where business and environmental interests can both benefit.

8.3 Creating SB 1547

In October 2015, a particularly bruising full session of the Oregon State Legislature saw the failure of a hotly debated statewide cap-and-trade program for carbon emissions, the failure of a "coal-to-clean" bill, the adoption of minor bills on solar energy and storage, and an active conflict between the petroleum industry and the environmental community over the state's low-carbon fuel standard. Environmental advocates were feeling stymied on progress to reduce carbon emissions. Similarly, Oregon's investor-owned utility community was searching for a path forward in the face of increasing state and federal policy efforts regarding carbon emissions in the electricity sector. The efforts of a newly founded organization would force utilities to action.

8.3.1 Utilities Face Disruption

In early October, a newly formed organization, Renew Oregon, filed four ballot measure petitions with the Oregon Secretary of State's office. With minor differences, the petitions all sought two key policy changes: (1) an increase in Oregon's RPS from 25% to 50% on an aggressive timeline starting in 2020 and (2) a requirement that Oregon electric utilities remove all coal-generated electricity from rates by 2030 or the date the coal-generating facility would be fully depreciated, whichever was earlier. The thrust of these concepts was straightforward: utilities should cease serving customers with coal power. This would increase the economic pressure on coal generation, possibly leading to additional unit closures. At the same time, by increasing fuel-free renewable energy sources in the marketplace, the higher RPS would undercut the price of coal power. This, too, would lead to the economic shutdown of coal-fired generating units. While these proposals sought to improve environmental outcomes, they did little for the other elements of sustainability, including economic and societal issues. Furthermore, the proposals did not integrate with existing utility resource planning, did not account for known retirements of coal fleet resources such as Boardman and the timing of replacement resources, and did not consider the cost impact of the regulations.

As the main investor-owned electric utilities in the state, and as companies that had sustained typically positive relationships with their customers, PGE and PacifiCorp immediately recognized the political threat posed by these petitions. Based on customer surveys by both companies over the previous decade, the utilities had long understood that customers want to move to more sustainable energy generation. This has led to policies and programs focused on environmental health and sustainability. PGE understood that if one or more of the petitions qualified for the ballot, there would be a significant possibility that voters would adopt the measures.

The issue was not that the companies opposed policies to increase renewable energy production or limit coal energy in the future; both must happen if the state is to meet its legislatively adopted emissions goals. The issue was that the measures were drafted in a way that would unnecessarily increase costs for electricity, resulting in negative consequences for the economy and PGE's customers.[3] The utilities agreed they could meet the measures' overall goals, but inflexible timelines and the lack of price and reliability protections[4] meant the utilities might have to oppose the measures themselves. In addition, there were other improvements that PGE wanted

[3] The proposed measures would have eliminated the value associated with banked Renewable Energy Credits (RECs), which, under the 2007 RPS, were allowed to be banked indefinitely to be used for compliance. Under the measure, the life of RECs would have been limited to 3 years. More costly to customers, however, was the requirement to increase renewable percentages even when not tied to planned or future coal resource retirements. Better alignment to those energy needs resulted in reduced overall cost to customers.

[4] Achieving a 50% renewable requirement with currently available variable technologies raised concerns about the ability to serve customer loads without the potential loss of service. The final language negotiated the ability to pause compliance with the standard should federal reliability standards be implicated.

to make if the RPS were to be significantly revised through legislation, as well as opportunities to address other topics, such as transportation electrification, community solar, and reliability provisions.

8.3.2 A Negotiating Workgroup Forms

Rather than engage in an expensive and divisive ballot campaign, the utilities invited Renew Oregon to help develop a compromise that would benefit the environment while addressing timing, price, and reliability concerns. Driven by legislative deadlines, the parties met during a compressed, 6-week schedule and announced their joint plan in January 2016.

The workgroup began meeting in November 2015 with a team of representatives from the utilities, environmental organizations, customer advocates, and renewables interests. The intention was to keep negotiations to groups that would bring a collaborative perspective to the table. Environmental groups known for more confrontational approaches were not invited into the work sessions, though all parties understood they were tracking the discussions and weighing in with other environmental organizations in the workgroup. Once the workgroup had agreed on a general framework, it began to explore the needs of individual renewable energy development interests and other environmental and equity organizations. By early December, the workgroup was ready to start drafting new language for the bill and develop cost impacts. Working off the framework agreement before drafting specific language kept wordsmithing to a minimum, although several sessions were held to work through the details. A small number of attorneys from the utilities and renewable development interests were dispatched to write detailed language with legislative bill drafters who work in the Oregon State Capitol.

Between November and January, key OPUC staff were generally kept apprised of the progress, and the workgroup sought OPUC's views on the substance of the negotiations. However, the workgroup did not include OPUC staff or commissioners in the initial negotiations, opting to bring in considerations heard from OPUC through informal channels. In hindsight, PGE and other workgroup participants acknowledge that greater engagement with OPUC during the workgroup process would have been beneficial, especially given the shortened timeframes for moving the bill through the legislature in the month-long short session. This experience highlights the need for clear communication and leaving more time for key stakeholders like the OPUC to weigh in throughout any such high-stakes negotiation.

After weeks of intensive negotiation, SB 1547, or the Oregon Clean Electricity and Coal Transition Plan, was created.[5] The bill was passed by the Senate on February 16, 2016, and by the House on March 1. Governor Kate Brown signed the bill into law on March 8 (Fig. 8.2).

---

[5] The bill was introduced to the Oregon House of Representatives as House Bill (HB) 4036. Due to the short legislative session, drafters had to amend a senate bill with the language contained in HB 4036.

Fig. 8.2 SB1547 Signing

8.3.3 Senate Bill 1547 Details

Oregon Senate Bill (SB) 1547 doubled the state's renewable standard, set firm timelines for transitioning the power delivered to customers away from coal generation, and promoted community solar, transportation electrification, and energy efficiency. While it was similar in scope to the ballot measure petitions, it differed in key attributes:

- The measures had contained an aggressive timeline for renewable energy increases, which would start in 2020 and ramp up very quickly. This timeline would have interfered with existing utility planning processes and been very expensive. SB 1547 called for a more modest increase, starting in 2025 and increasing at a slower pace.
- The measures significantly reduced the value of banked Renewable Energy Credits (RECs) that had been held by utilities for future compliance, and RECs would have expired after three years. SB 1547 retained the value of banked RECs and extended the life of new RECs to five years.
- The coal phaseout was extended by a few years to take into account the unique ownership structure of the Colstrip Steam Electric Station and the implications of that ownership structure for a single-state utility like PGE. SB 1547 permits PGE to ask OPUC to allow Colstrip to be included in rates until 2035 if necessary. The measures would have removed all coal-generated energy from rates by 2030.

- Recognizing the potential for reliability issues, the utilities added language to allow for a pause in compliance if higher penetration of variable energy resources like wind might impact reliability but only in a fashion that did not absolve the requirement to ultimately comply with the increased RPS.

The utilities argued that larger increases in the RPS tiers were best reserved for years when existing generation was expected to exit utility portfolios. This allowed utility planning, as well as social and economic impacts, to be considered when determining how best to replace existing resources, rather than requiring large increases in renewable energy when the additional energy was not actually needed. PGE also advocated for keeping the 2020 renewable target the same as under the 2007 RPS, since utilities were already developing integrated resource plans that included 2020 in the planning horizon.

For comparison, the ballot measures would have called for:

- A 22% requirement for renewables in 2020 (up from 20%).
- A 30% requirement in 2025.
- A 40% requirement in 2030.

Under SB 1547, the new standards include:

- A 27% requirement in 2025 (a 2% increase from the former requirement).
- A 35% requirement in 2030.

The ballot measures and SB 1547 both called for a 45% requirement in 2035.

The RPS adopted in 2007 allowed unlimited banking of RECs. Utilities had developed sizeable banks of RECs, thanks to early compliance, and anticipated using them to smooth the timing of energy facility installation. The ballot measures would have turned these RECs into certificates with a limited, 3-year life and allowed for the creation of 3-year RECs. In other words, the REC would expire after 3 years and companies could no longer use them to meet compliance requirements. The workgroup modified REC life provisions to 5 years while retaining the unlimited life for RECs already banked. This provision alone ensured that the early renewable energy investments by PGE's customers would continue to provide future compliance flexibility. The workgroup also drafted provisions to encourage the early siting of renewable energy resources by allowing facilities sited within the first 5 years of the adoption of SB 1547 to create unlimited life RECs for a certain period of time.

Finally, the ballot measure drafters did not fully understand the complicated ownership structure of the Colstrip facility at the beginning of this process. Both PGE and PacifiCorp, the other large utility affected by this legislation, own a portion of the Colstrip coal facility located in Montana. However, the facility is owned by a total of six entities, including two Washington utilities, one Montana utility, and a merchant owner. While it is understood that the plant has a real lifetime, it is not possible for one owner to demand that the facility close. Therefore, the negotiating group drafted a provision giving a utility like PGE with load only in Oregon the option of extending by 5 years, to 2035, the obligation to remove the costs and benefits of the plant from customer rates.

8.4 Benefits to the Environment

While negotiating SB 1547, it was critical to ensure that the proposal met environmental goals. The reality is that utility companies and environmental advocates in Oregon have not disagreed on the need to reduce the emissions that affect climate change. As noted, PGE has worked cooperatively for many years, at both the federal and state level, to support policies that reduce emissions. PGE has questioned environmental policies on the principal that reductions in carbon emissions must come at a reasonable pace, with acceptable rate impacts to customers. This approach helps protect customers from rate shock and avoids imposing costs that are not necessary.

When Oregon adopted ambitious, economy-wide emissions reduction goals in 2007, the end target was a 75% reduction from 1990 emission levels by the year 2050. While the goals were never intended to directly translate to a specific sector or company, it was important to ensure that SB 1547 would put the electricity sector on a path to reductions consistent with the state-wide goals. PGE predicted that the legislation would place the company close to that imputed reduction at least through 2040 on a straight line to the goals in 2050 (Fig. 8.3).[6] Should PGE achieve those reductions with the changes made in the legislation, PGE believes the rate impacts on customers will be manageable.

Similarly, PacifiCorp estimated that the difference in emissions between the proposed ballot measures and SB 1547 was negligible over the long term (Fig. 8.4) [6].

With this plan, the two main investor-owned utilities, which serve nearly 70% of the state's electricity load, would see combined reductions through 2035 below the straight-line 2050 goal trajectory as estimated by the chair of the Oregon Global Warming Commission.

8.5 Benefits to the Economy

One critical aspect of drafting a sustainable piece of legislation was lowering the end costs for utility customers while pursuing environmental benefits.

While strong energy-efficiency efforts have resulted in residents using less energy per capita than in 38 other states, [7, 8][7] Oregon's investor-owned utilities remain concerned about low- and moderate-income residential customers who struggle to pay their bills. In 2015, 15.4% of Oregonians, higher than the national average [9], lived below the federal poverty level. A significant percentage of

[6] Since this initial projection, PGE has updated this graph and now predicts emissions to be even lower, due to actions taken in PGE's 2016 Integrated Resource Plan.

[7] Oregon regularly scores in the top ten, and often the top five, of states for energy efficiency as ranked by the American Council for an Energy Efficient Economy.

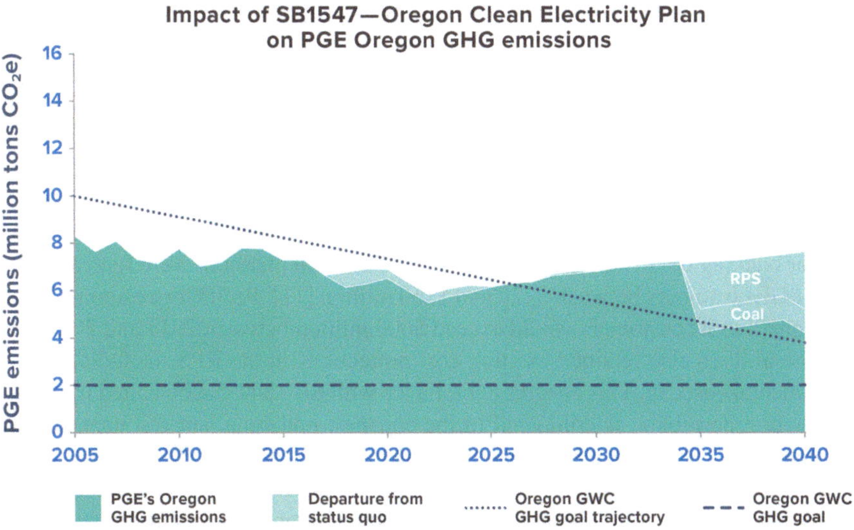

Fig. 8.3 The carbon emission savings from SB 1547 are projected to align with goals set by the state in 2007. Underlying emissions projections may be updated, as the graph is based on data available at the time of SB 1547 negotiations

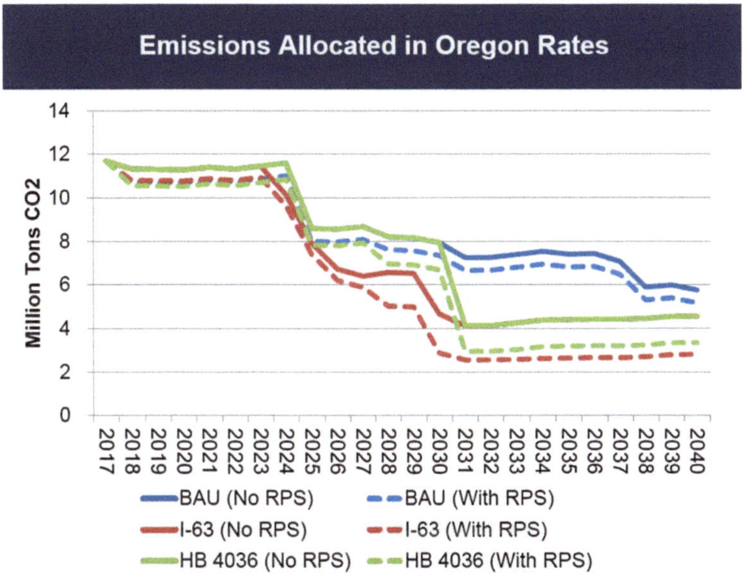

Fig. 8.4 Based on PacifiCorp estimates, the difference in emissions between the proposed ballot measures (labeled here as I-63) and SB 1547 (labeled here as HB 4036) was negligible over the long term

Oregonians live above the poverty level but struggle to choose between paying rent or utilities each month [10].[8]

As a result, the utility companies were concerned about additional costs to customers, which could come as a result of interference with existing utility planning processes, the overbuilding of renewable resources in advance of need, the elimination of cost-savings mechanisms from the RPS, and the advancement of coal shutdown in a way that ignored out-of-state co-owners. The changes between the ballot measure petitions as filed and SB 1547 relating to the timing of RPS increases, longer banking for RECs, and the extension of the coal phaseout were forecasted to save PGE customers between $220 and $360 million [11]. PacifiCorp estimated that the benefit to its customers would exceed $600 million between 2017 and 2030 [6].

The utilities also retained existing cost protections in the RPS, including a 4% incremental cost cap. This ensures compliance with SB 1547 can be halted if investments in renewables, as compared to the cheapest noncompliant resource, would exceed 4% of the revenue requirement of the regulated utility.

Ultimately, the utilities were able to ensure the environmental community's end goal—the reduction of emissions consistent with the state's targets in the electricity sector—was met, if not exactly at the same pace. At the same time, utilities ensured that careful timing, cost protections, utility planning, and system reliability were not ignored. These trade-offs offer an excellent example of how to support sustainable choices by securing benefits across multiple sectors, rather than focusing on one to the detriment of another. Costs to PGE customers will still gradually rise under SB 1547 as the utility acquires renewable energy generation, but thanks to the new timeline for emissions reductions, they will not rise unnecessarily or as fast as they would have under the ballot measures.

8.6 Benefits to Society

Oregon's 2007 RPS measure did not contain any mechanism to limit requirements for additional renewables if the reliability of the grid was at risk. While much of the new renewable energy would likely come from variable energy resources like wind, the existing utility grid in the Pacific Northwest, with its reliance on hydroelectricity, could likely absorb much of the 25% renewable requirement. That is, the flexibility of being able to store water and generate energy with hydroelectric turbines when the wind was not blowing allowed for balancing significant wind generation with other generation resources.

Given the new renewable requirement, many stakeholders were uncertain as to how utilities would actually accomplish the integration of 50% variable energy into the grid while still keeping the lights on. While technologies change over a span of

[8] Oregon is slightly above the national average for the population living below 200% of the federal poverty line.

20 years and technologies like battery storage will continue to decline in cost, there are significant unknowns about timing and technology. As a result, many stakeholders were concerned about the potential for reliability impacts.

The workgroup took this opportunity to adopt reliability as part of the renewable standard. This feature is unique to Oregon. The provision allows the utility or OPUC on its own initiative to determine that compliance with increasing renewable penetration may implicate reliability standards. If the commission agrees that there may be reliability issues, utilities may pause additional compliance, develop a plan to restore predicted reliability, and implement the plan without being penalized for failing to comply with the RPS during the pause.

8.7 Other Emissions Reduction Activities in SB 1547

Oregon utilities have been working toward transportation electrification since before the release of the first mass-market electric vehicle by Nissan in 2011. Environmental advocates and utilities have recognized that the only way to effectively reduce carbon emissions from the transportation sector—and for that matter, to reduce emissions consistent with the state's goals—is to shift transportation emissions to the electricity sector and gradually reduce emissions from electricity while maintaining reliability and affordability.

However, regardless of potential benefits to the environment and utility system, previous OPUC decisions favored the competitive market, not regulated utilities, as the primary provider of transportation electrification infrastructure. In 2012, OPUC issued an order [12] that strictly limited investments to geographic areas in which it could be shown that third parties would not invest and customers would experience a compelling benefit for a specific investment. That burden was significant enough that it seriously limited utility involvement and investment in charging infrastructure.

This created a hurdle for utility transportation electrification plans. To encourage the purchase of electric vehicles, charging infrastructure is needed to overcome concerns about limited range. But investment in this infrastructure would come only if competitive providers had a business model that worked—and that required more electric vehicles. This "chicken and egg" problem could be resolved by a market actor with access to capital and the ability to take a longer-term view than most competitive entities—which PGE saw as an appropriate role for an investor-owned utility. However, PGE could not act without legislation that redefined the state's policies regarding transportation electrification and the role of the utility to support such efforts.

Environmental stakeholders advocated for addressing these issues as part of SB 1547. The workgroup drafted a requirement for utilities to file transportation electrification plans with OPUC that allowed utilities to recover the costs of investments in infrastructure or incentives to encourage electrification. The group provided further considerations for OPUC in allowing the investment to be paid for by utility customers.

On a different matter, the workgroup agreed on legislative language codifying the existing practice for investor-owned utilities to plan for and acquire all cost-effective energy efficiency. When it comes to decarbonizing the electricity sector, it is clear that not only must the electricity itself come from cleaner sources, but it must also be used efficiently. Utilities must also plan for cost-effective demand response to lower the need for reserve energy sources in peak demand. While utilities have been doing this for many years, this language made the acquisition a requirement. As the cheapest kilowatt is the one you don't have to buy, PGE believes this language will ultimately reduce costs to customers on a long-term basis.

Another portion of SB 1547 established a community solar program. Similar to programs already seen in other states, Oregon will allow customers to receive credit for energy produced by a solar facility that is not attached to their residence or commercial business. As of 2017, the state is still developing rules for this program and a "resource value of solar" to compensate customers who subscribe to community solar projects. Apartment dwellers, people who own homes with shaded rooftops or poor orientation, and people who simply do not wish to own their own solar facility can join with individuals in supporting solar development.

Finally, in lobbying SB 1547 through the legislature, its opponents made an argument that natural gas generation resources, cleaner than coal by far, would be a better investment for customers. The workgroup showed that in the long term, fuel-free renewable resources were best when considering cost, environmental benefits, and societal impact (Fig. 8.5).

8.8 Conclusion

As a result of the collaborative engagement that led to SB 1547, Oregon's electricity sector will be on track to meet the state's greenhouse gas reduction goals, reducing emissions by 75% below 1990 levels by 2050. The legislation included important provisions to keep utility prices manageable and ensure reliability for all businesses and customers. The environmental benefits were clear as well, reducing carbon dioxide pollution across the western states by an estimated 30 million metric tons—the equivalent of taking 6.4 million cars off the roads. Once the legislation's renewable energy targets are met in 2040, the annual carbon dioxide emissions in the western states will be reduced by four million metric tons from current levels. Investments PGE has made over the past decade are already preventing nearly 1.6 million tons of carbon emissions per year. Combined with the planned closure of the Boardman facility, PGE is on track to achieve another 2.2 million tons of carbon reductions annually through 2021 due to energy efficiency measures and renewable power development.

In conclusion, SB 1547 will achieve virtually the same environmental benefits as intended with the original ballot measure petitions. What is truly notable is that in a

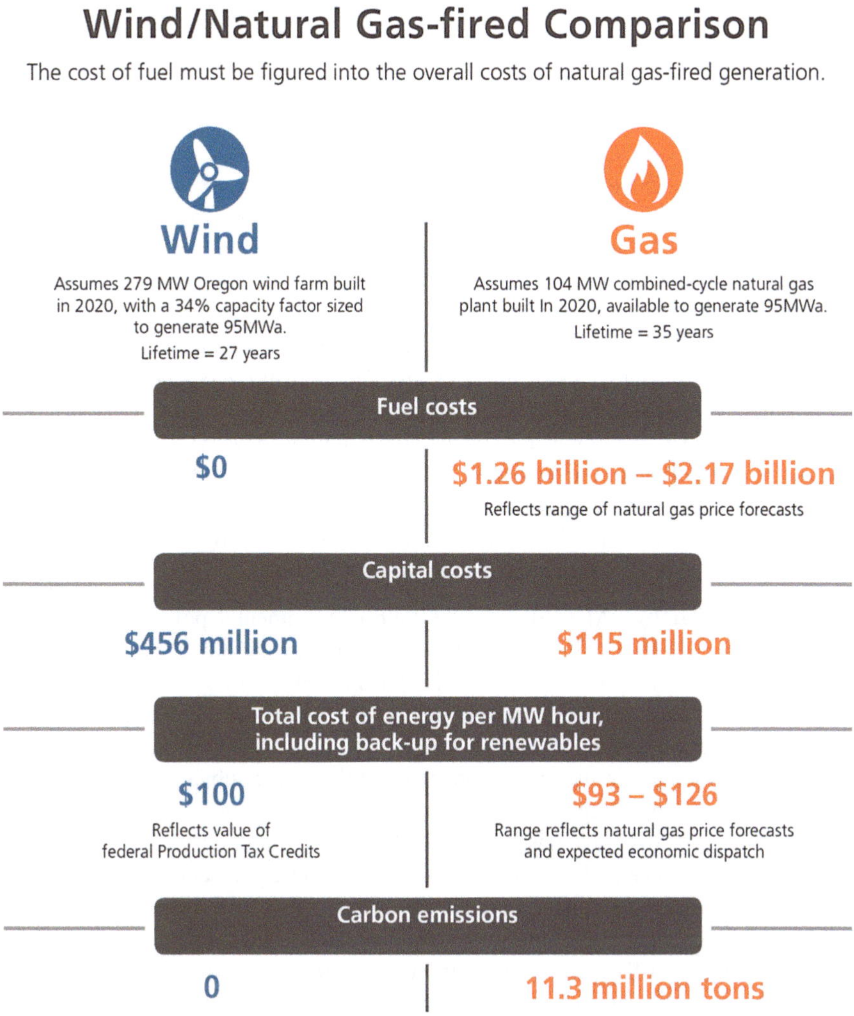

Fig. 8.5 In the long term, fuel-free renewable resources have the best results for cost, environmental benefits, and societal impact

matter of 3 months, a diverse group of advocates, businesses, and utilities managed to fast-track a historic negotiation that led to a responsible path toward carbon reduction. At a manageable 1.5% annual rate impact to customers, a significantly lower increase than with the alternative plan, SB 1547 offered practical trade-offs that ensured environmental, economic, and social sustainability.

Author Profiles

Dave Robertson Robertson oversees PGE's corporate communications, government relations, environmental policy, and corporate social responsibility groups. He joined PGE in 2004. Before taking on the role of vice president in 2009, Robertson served 5 years as director of PGE's Government Affairs department.

Robertson has 25 years of experience in the energy policy area. Before joining PGE, he served as director of government relations (West) for Pacific Gas and Electric's National Energy Group. Robertson was a member of US Sen. Mark Hatfield's legislative policy staff in both Washington, D.C., and Oregon and was a consultant at Robertson Grosswiler & Co. as well.

Robertson is the incoming chair of the Portland Business Alliance and a member of the Oregon Business Association Board and Industry Association Executive Committee. Previously, he served as a member of the Governor's Transportation Vision Panel and as an Oregon appointee to the Columbia River Gorge Commission. He is a Senior Fellow of the American Leadership Forum of Oregon and completed the Utility Executive Course at the University of Idaho.

He earned a bachelor's degree in political science from Willamette University and is a graduate of Parkrose High School in East Portland.

Brendan McCarthy McCarthy is the state environmental policy manager for Portland General Electric. In his role, he develops policy initiatives, performs stakeholder relations, monitors issues, and advocates for PGE in legislative and regulatory contexts in Oregon, Montana, and the Pacific Northwest. He was instrumental in the adoption of Oregon's original renewable energy standard in 2007, helped negotiate the increase in the renewable standard that is the subject of this article in 2016, helped design Oregon's emissions performance standard, and has shaped various solar initiatives. He has been with PGE since March 2006 and before that worked for the Office of Legislative Counsel in the Oregon Legislative Assembly and focused on drafting legislation in the environmental and natural resources areas. He also has experience working on local, state, and congressional political campaigns. He holds a BS degree from Penn State University and a JD degree from the Northwestern School of Law of Lewis and Clark College in Portland.

Company Profile

Portland General Electric, headquartered in Portland, Oregon, is a fully integrated energy company that serves approximately 875,000 residential, commercial, and industrial customers in 51 cities and covering 4,000 square miles in Oregon. For more than 125 years, PGE has been carrying out its mission to consistently deliver safe, reliable, and responsibly generated electricity. This is done through a generation mix that includes an increasing amount of renewables—wind, hydro, and solar—as well as natural gas, coal, and purchased power. As a member of the western Energy Imbalance Market, PGE is now part of an automated system that integrates electricity generation across six states, better integrating renewable energy, managing the variations of customers' load, and capturing the most cost-effective resources. PGE delivered nearly 19 million megawatts of power in 2017 with 99.98% reliability. PGE's $6.7 billion in net utility plant assets includes 17 plants—2 wind, 7 hydro, and 8 thermal, plus a network of solar facilities—located in Oregon, Washington, and Montana. Together with its customers, PGE has the no. 1 voluntary renewable energy program in the United States. PGE is committed to delivering clean energy solutions, affordably and seamlessly, through an integrated, secure, smart energy grid. For more information visit PortlandGeneral.com.

References

1. "About the Sierra Club," Sierra Club, https://oregon2.sierraclub.org/chapter/about. Accessed 5/22/2018.
2. "More than 800 Oregon Business Support the Clean Energy Jobs Bill: Across the State and Sectors," Renew Oregon, 27 January 2017, www.reneworegon.org/700_oregon_businesses_support_the_clean_energy_jobs_bill. Accessed 22 May 2018.
3. D. Olick, "Why people are flocking to Oregon," CNBC, 5 January 2016, www.cnbc.com/2016/01/05/why-people-are-flocking-to-oregon.html Accessed 22 May 2018.
4. "Summary: The Environment 2013 Oregon Values & Beliefs Study," OR Values and Beliefs Project, http://oregonvaluesproject.org/ovp-content/uploads/2013/10/OVB_Environment_Summary.pdf. Accessed 22 May 2018.
5. *Utility Trusted Brand & Customer Engagement: Residential*, Marketing Strategies International (MSI) 2017, https://landing.marketstrategies.com/utility-trusted-brand-and-customer-engagement-residential-study-2018). Accessed 22 May 2018.
6. "Pacific Power – HB 4036 Analysis, Testimony of Scott Bolton, February 2016, to the House Committee on Energy and Environment, https://olis.leg.state.or.us/liz/2016R1/Downloads/CommitteeMeetingDocument/83847. Accessed 22 May 2018.
7. "Rankings: Total Energy Consumed per Capita, 2015 (Million Btu)," United States Energy Information Administration, www.eia.gov/state/rankings/?sid=OR. Accessed 22 May 2018.
8. "State and Local Policy Database," American Council for an Energy-Efficient Economy, https://database.aceee.org/state-scorecard-rank. Accessed 22 May 2018.
9. "Oregon's Poor Still Feel Weight of the Great Recession," Oregon Center for Public Policy fact sheet, 12 October 2016, www.ocpp.org/2016/10/12/oregons-poor-still-feel-great-recession/. Accessed 22 May 2018.
10. "Distribution of the Total Population by Federal Poverty Level (above and below 200% FPL)," Henry J. Kaiser Family Foundation, www.kff.org/other/state-indicator/population-up-to-200-fpl/?currentTimeframe=0&sortModel=%7B%22colId%22:%22Location%22,%22sort%22:%22asc%22%7D. Accessed 22 May 2018.
11. "HB 4036 Oregon Clean Electricity & Coal Transition Plan," Testimony of Varner Seaman, February 2016, to the House Committee on Energy and Environment, (https://olis.leg.state.or.us/liz/2016R1/Downloads/CommitteeMeetingDocument/83714). Accessed 22 May 2018.
12. "Investigation of matters related to Electric Vehicle Charging," Before the Public Utility Commission of Oregon, Order No. 12013, 19 January 2012, https://apps.puc.state.or.us/orders/2012ords/12-013.pdf. Accessed 22 May 2018.

Chapter 9
Silicon Valley Customer Expectations: Resilience and Innovation in a Changing Climate

Marianna Grossman

Abstract Silicon Valley is often a harbinger of change, including in the world of electricity where information technology is playing a growing role. This chapter explores how companies and other large institutional customers are taking their energy production and management into their own hands in order to control costs and lower greenhouse gas emissions. Large customers are joining together to demand renewable energy at a massive scale. Utility executives and regulators are grappling with how to change their business models and take action to keep customers satisfied. In the San Francisco Bay Area, utilities like Pacific Gas and Electric (PG&E) are providing transmission infrastructure and billing services to support Community Choice Energy cooperatives, such as Silicon Valley Clean Energy. New green tariffs and other models for delivering renewable electricity are emerging across the country.

9.1 First Steps: Conservation and Efficiency

In the early days of sustainability, many companies and institutions focused on saving money by reducing energy use through conservation (using less), energy efficiency (wasting less), and direct access to negotiate contracts for sourcing lower-priced energy. As scientific data about climate impacts and weather changes have become more alarming and incontrovertible, more large energy users are viewing the challenge of energy as an important way that they can contribute to a more resilient and less volatile future. Large energy users such as Apple, Facebook, Google, HP Inc., VMware, Stanford University, and cities such as Mountain View, Palo Alto, and San Francisco have committed to using 100% carbon-free energy and to reducing their overall energy use, especially from fossil fuels. As large energy

M. Grossman (✉)
Minerva Ventures, Mountain View, CA, USA
e-mail: mgrossman@minervaventures.com

J. Fox, M. Scott (eds.), *Sustainable Electricity II*,
https://doi.org/10.1007/978-3-319-95696-1_9

users found ways to drive down demand and increase efficiency, they began to look for a stable supply of low cost, clean, renewable energy, in many cases making the investments to produce it or source it directly.

9.2 Next Steps: Renewable Energy and Smart Technology

It has not been easy for large energy users to switch away from fossil fuel-based electricity. This chapter will distill the key strategies and tactics that some large energy users have developed to work with utilities and other energy providers. They have added smart technology and data-driven approaches to reduce their overall energy use and to switch to lower-carbon energy sources: saving money, reducing problems of uneven energy quality, and meeting their climate protection goals ahead of schedule. These customers require detailed data on energy quality and use so that they can better manage both supply and demand, using a wide range of technology from smart meters and machine learning to sensors. They have realized additional benefits such as reducing air pollution and noise due to switching from vehicles with internal combustion engines to electric propulsion. Employers also find that pro-environment policies and practices help them attract and retain employees in a competitive market for tech talent.

Through rare interviews, company-approved stories, and detailed case studies, this chapter will explore the expectations of companies in the Silicon Valley—a region that has been known to lead culture change worldwide.

9.3 Case Studies

The five following case studies—of VMware, HP Inc., Facebook, Google, and Stanford University—highlight some of the challenges faced by these companies in combatting climate change through switching to renewable energy and some of the solutions employed.

9.3.1 VMware[1]

> "With the announcement that the US will withdraw from the Paris Climate Agreement, I wanted to reiterate VMware's absolute commitment to operating our business in a responsible and sustainable manner. From our commitment to green building practices around the world to the order of magnitude our technology delivers in terms of energy and carbon emission reductions in IT, VMware has been on the leading edge of positive environmental impact

[1]Case study based on dialogue with and participation from Nicola Acutt, VP Sustainability Strategy, Office of the CTO, and Natasha Tuck, Senior Sustainability Manager of VMware in Dec 2017 and Jan 2018.

since our inception... VMware will continue to be vigilant and innovative in driving a net positive future for VMware and our customers. In 2016, we announced our goal to achieve carbon neutral emissions by 2020 and to advance our commitment to use 100 percent renewable energy. We stand by that commitment because it's the right thing to do for our employees, our company, our customers and our planet." Pat Gelsinger, CEO, VMware [1]

VMware Products Help Customers Use Less Energy
VMware is a fast-growing company, founded in 1998, that creates and delivers software to increase the efficiency of compute servers, as well as storage and networking systems. IDC's 2017 analysis [2] documents that VMware vSphere virtualization has a significant, tangible, and net positive effect on lowering the size and scope of infrastructure and, as a direct result, reducing the carbon intensity of VMware's customers' IT environments. VMware enables companies to aggressively reduce their carbon footprint through the modernization of data centers and public cloud integration. VMware's virtual infrastructure products allow users to reduce computing energy use by up to 80%, which has resulted in energy savings of billions of kilowatt hours of electricity for VMware customers according to IDC's 2016 study [3].

Challenge 1: Energy Efficiency for a Diverse Real Estate Portfolio
As a growing tech company, VMware has a global portfolio of leased office space, and as a result, there is no "cookie-cutter" approach to energy reductions or renewable energy procurement. Sustainability presents the ongoing question of whether to set a target because you *know* you have to get there or to figure out *how* you're going to get there first. VMware's motto is making the seemingly impossible possible, so they plan to get there either way. With a commitment to science-based targets and carbon management, VMware's Real Estate and Workplace team evaluates opportunities to reduce both VMware's carbon footprint and their operating costs.

Solution 1: VMware Makes Operational Investments via Roadmap
In an effort to accelerate their progress toward reducing their overall energy consumption, they have conducted energy audits across their sites. Based on opportunities uncovered by the audits, VMware has developed a 3-year roadmap that staggers energy-saving projects in order to absorb them financially with the goal of reducing energy use and achieving 100% renewable energy by 2020. Examples of operational investments that help achieve these goals include numerous infrastructure projects—from LED lighting upgrades throughout their garages at their headquarter campus in Palo Alto to LEED® Platinum certification for their site in Bangalore, India. In 2016, VMware reduced their emissions intensity by 15%, significantly surpassing their 10% target. VMware achieved 8% financial growth while reducing scope 2 emissions by 4%, demonstrating that business fundamentals and environmental responsibility can go hand in hand.

Challenge 2: Sourcing Renewable Energy
With over 116 global and diverse locations served by many utilities, sourcing renewable energy is difficult. Rather than working on their own with each utility, could they join with others to access renewable energy for their sites?

Solution 2A: Partnerships for Renewable Energy and Verified Offsets
To better understand the renewable energy landscape and identify opportunities to procure local green power, VMware has partnered with various initiatives that are leading the way toward a low-carbon future, including the Corporate Renewable Energy Buyers' Principles [4]. VMware also supports RE100 [5], another powerful force in the market. Officially launched at Climate Week NYC in 2014, RE100 is a global collaborative representing businesses committed to using 100% renewable electricity in partnership with The Climate Group[2] and CDP.[3] Through these partnerships, VMware joins some of the most influential companies committed to accelerating the transformation of the global energy market and aiding the transition toward a low-carbon economy.

VMware has aligned their strategy with RE100 and the GHG Protocol,[4] which has helped them develop a thoughtful strategy for the procurement of energy instruments and low-carbon sustainable development projects (aka carbon offsets). They are taking an incremental approach with a plan to ramp up their procurement of renewable energy instruments and carbon offsets each year through 2020, integrating a price on carbon into their business over time. This also creates a platform against which internal or external options can be measured and adjusted. In 2016, for the second consecutive year, they reduced to zero their scope 2 emissions in the USA with the purchase of 14,550 metric tons (MT) CO_2e (or 30,000 MWh) Renewable Energy Credits (RECs). Currently, VMware is procuring 72% renewable energy globally and is making solid progress toward their 100% renewable energy and carbon neutrality goals for 2020.

Solution 2B: VMware Uses Clean Power at its Data Center
VMware designed and constructed a LEED® Platinum certified data center in East Wenatchee, Washington, to be one of the greenest, most sustainably engineered, and energy-efficient data centers in the world. They selected this location on the Columbia River to take advantage of the 100% hydroelectric power provided by the local utility as well as the mild climate, which allows for free air cooling most of the year. This facility features breakthrough containment methodology and efficient virtualization, which combine to make it a future-focused model of sustainability, efficiency, and cost savings. They have achieved a power utilization efficiency (PUE) of 1.25, well below the industry average of 2.2. They are saving more than 39,000 MT CO_2e annually and have deployed virtualization with 100% clean, renewable energy. Their on-site gray water system for water closets reduces potable water use by 74%.

---

[2]The Climate Group https://www.theclimategroup.org/ brings together business and government to take action for the climate.

[3]CDP is a nonprofit organization that offers environmental reporting services (https://www.cdp.net/en).

[4]The GHG Protocol https://ghgprotocol.org/ was developed jointly by the World Resources Institute (WRI) and the World Business Council for Sustainable Development (WBCSD).

Fig. 9.1 VMware HQ campus with solar panels, Palo Alto CA 2017. (Source: VMware)

Solution 2C: Sourcing Green Power Directly
In 2015, VMware made its first voluntary purchase of green power, purchasing 60,000 MWh of American Wind™ RECs, which offset more than 100% of their US operations. They have continued to procure 100% renewable energy in the USA and, for the first time in 2017, achieved 100% renewable energy in their Europe, Middle East, and Africa (EMEA) region. On-site solar opportunities are limited in office locations the company does not own. However, in the company-owned Palo Alto headquarters, they have installed an on-site solar system through a power purchase agreement (PPA) for approximately 1 MW and are covering the roofs of two new buildings that will be completed in 2018 with another 0.5 MW of solar (Fig. 9.1). In addition, VMware purchases carbon-free electricity from the City of Palo Alto Utilities.

VMware understands that their footprint extends beyond their walls and, as a result, has furthered their analysis and reporting of scope 3 emissions. In 2017, they received third-party assurance for their scope 3 emissions related to business travel for their CDP report [6], and they have started down the path of calculating other scope 3 emissions.

Solution 3: VMware Supports Electrifying Transportation
VMware is a member of the US Department of Energy's Workplace Charging Challenge, a national partnership program aiming to increase the number of employers offering workplace electric vehicle (EV) charging. Ten percent of VMware headquarters employees are registered EV users. Currently, there are 72 charging ports in VMware's parking garages, where VMware employees used electricity in place of 15,000 gallons of gasoline in 2017. VMware seeks to provide additional

opportunities to support their employees in living a sustainable lifestyle personally and professionally. Their goal is to pave the way toward a low-carbon economy.

Driving Force: Collaboration

There are many approaches to meeting ambitious goals like RE100, and the solutions available are constantly evolving. For this reason, VMware believes that it is important to share thinking, learning, and insight. The bottom line is that electricity customers need to demonstrate more demand for renewable energy to ensure the expansion of renewable energy capacity. This demand can be created by policy-makers through compliance-based renewable energy targets or through voluntary actions taken by businesses and individuals. VMware is taking responsibility for their impact and joining other corporate leaders through voluntary action and sharing their experience. Getting to 100% renewable energy is a bold and ambitious goal that inspires VMware employees to innovate and to create new ways of doing business—putting back more than the enterprise takes, and ultimately having a net positive impact on the world.

> "VMware has always been driven by the belief that the seemingly impossible is possible. The company's commitment to sustainability is reflected in everything it does—from the products it creates to its everyday operations. VMware has been instrumental in enabling cleaner compute and energy reductions across its vast customer base—positively impacting businesses in every sector. VMware's virtual infrastructure products allow users to reduce computing energy use by up to 80 percent, which has resulted in energy savings of billions of kilowatt hours of electricity for VMware customers. VMware knows there is a lot of work to be done, but the company is excited about the possibilities for transformation and innovation that lie ahead." Nicola Acutt, VP Sustainability Strategy Strategy, Office of the CTO. VMware Commits to Carbon Neutral Emissions by 2020. September 20, 2016. https://www.vmware.com/radius/vmware-commits-carbon-neutral-emissions-2020/

9.3.2 *HP Inc.*[5]

> "HP's goal is to achieve 100% renewable electricity. Our interim goals are to achieve a 25% reduction of carbon emissions and to be 15% more water efficient, based on 2015 levels, by 2025. Sustainability is core to our values at HP and is demonstrated in how we do business. Achieving these goals is difficult because of the complexity of our global operations, but our management, our customers and our employees inspire us to do more." Mary Curtiss, Global Head of Energy and Sustainability – Corporate Real Estate at HP, Inc.[6]

Multiple Driving Forces for Sustainability and Green Energy

In November 2015, Hewlett Packard Company separated into two companies: HPE (Hewlett Packard Enterprise) and HP Inc., the personal systems and printer side of the business that also includes HP Labs. HP executives and its board of directors are committed to environmental sustainability. CEO Dion Weisler is from Australia, so he is keenly attuned to the risk of water scarcity and the importance of addressing

[5] This case study is based on a personal interview with and additional material from Mary Curtiss, Global Head of Energy and Sustainability—Corporate Real Estate at HP, November 2017 and additional information from the company website. It has been reviewed and approved by HP, Inc.

[6] Personal Communication. November 2017

and responding to climate change. HP policy is driven by science, and HP is a member of the Business Environmental Leadership Council whose members declared their support of the United Nations Paris Climate Accord.

> "At HP, we're reinventing for a better world. Sustainability guides every aspect of our business, fuels our innovation and growth, and works to enable everyone, everywhere to thrive. It is the right thing to do for our business, for society, and for the long-term future of our planet…
>
> We're encouraged to see governments, corporations, organizations, and individuals supporting the United Nations Sustainable Development Goals (SDGs). In 2016, HP made meaningful progress on 15 of the 17 SDGs, and set bold new targets in key areas where we can have the greatest impact—including two major goals to drastically reduce greenhouse gas emissions in our operations and supply chain…" Dion Weisler, President and Chief Executive Officer, HP Inc., 2016 Sustainability Report [7]

HP's motivation to institute renewable energy and energy efficiency in operations is further inspired by customers who expect suppliers of electronic products to demonstrate compliance with the Electronic Products Environmental Assessment Tool (EPEAT) certification[7] and other measures of environmental performance. EPEAT certification is often included in request for proposal (RFP) specifications from key customers. This facilitates alignment between the sales and operations organizations in building a business case for environmental performance through cost savings, top-line growth, and bottom-line efficiency. In addition, employees are eager to work at companies with outstanding sustainability performance; thus, environmental factors become an incentive for hiring, employee engagement, and retention. Progress on environmental initiatives helps build positive morale and enthusiasm. Reliable long-term pricing for electricity, without the volatility of fossil fuel-based energy prices, helps HP keep energy costs under control.

Challenge 1: Finding Sources of Renewable Energy Across a Global Real Estate Portfolio

HP has global manufacturing facilities and 20 million square feet of office space in more than 60 countries. Outside of its Palo Alto headquarters, much of that office space is leased. There is a mismatch of incentives between commercial property owners and tenants that makes it difficult to meet ambitious sustainable energy goals.

Solution 1A: Partner with Others to Meet Renewable Energy Goals

In 2016, the renewable electricity HP generated on-site or purchased accounted for 15% of their total consumption. HP is a member of RE100 and has committed to a goal of 100% renewable electricity with an interim goal of 40% by 2020. Their overarching operations goal is to reduce scope 1 and scope 2 GHG emissions from global operations by 25% by 2025 compared to 2015 levels.

Solution 1B: Build Green Offices and Purchase 100% Renewable Electricity

HP has been working systematically to find renewable energy sources for its many sites around the world. For example, HP opened its new Singapore campus in December 2017 that includes its first Smart Manufacturing Applications and Research Center (SMARC), where advanced manufacturing solutions will be developed and tested, before being used to upgrade HP's operations across the globe. They are about to embark on a PPA for 20 years. This will be one of the largest PPAs in Singapore.

[7] For more information on EPEAT, see http://greenelectronicscouncil.org/epeat/epeat-overview/

In the USA, HP is working with a developer to build a new site in Houston, Texas, for 2400 employees. They plan to purchase 100% green energy. The move to a new campus was catalyzed by damage to their old site from 2 years of unprecedented flooding, including from Hurricane Harvey in 2017. Such flooding is an example of costs due to intensification of severe weather events caused by climate change, which can jeopardize investments in on-site renewables. Local energy generation and smart microgrids can increase resilience to extreme weather events, permitting business continuity while others wait for power to be restored as the traditional energy grid is rebuilt after disasters.

Challenge 2: Need Adequate, Detailed Energy Data
The lack of submetering at many sites makes it harder to target energy efficiency measures. For example, there may not be submeters for chillers to determine optimal efficiency and peak load management. This type of detailed data could be used to determine if solutions such as heat pumps could reduce energy use for building and data center heating and cooling.

Solution 2: Work with Utilities and Real Estate Managers to Provide Submetering and Detailed Energy Data
HP is working with the City of Palo Alto Utilities to better understand detailed energy usage at their headquarters. They will use this data to drive changes, such as upgrading chillers and other energy-intensive systems. They work with their corporate real estate management company to upgrade other sites around the world.

Driving Force: Corporate Resilience
HP combines committed leadership, customers who require energy-efficient and environmentally friendly products, employee engagement, and cost-management goals to set and achieve the aspirational goal of 100% electricity from renewable sources.

9.3.3 Facebook[8]

> "[The U.S.] Withdrawing from the Paris climate agreement is bad for the environment, bad for the economy, and it puts our children's future at risk.
>
> For our part, we've committed that every new data center we build will be powered by 100% renewable energy.
>
> Stopping climate change is something we can only do as a global community, and we have to act together before it's too late." Mark Zuckerberg, CEO, Facebook, June 1, 2017 [8]

Challenge 1: Improving Performance of Services, Data Centers, and Offices
Running a global social media platform requires large amounts of energy to power computers, data centers, and offices. The first goal is to reduce energy demands through efficiency and better design of systems and facilities.

[8]This case study is based on email correspondence between Nov 2017 and March 2018, company white papers and material from the Facebook website. It has been reviewed and approved for publication.

Solution 1: Optimizing Facebook Data Centers and Offices
Facebook data centers are LEED® Gold certified, powered by hyper-efficient hardware, and cooled using outdoor air and evaporative cooling instead of energy-intensive air-conditioning units. Their data centers are designed to be water efficient [9] and use 50% less water than typical data centers. They reuse water as much as possible and use water-efficient fixtures. They install low-water landscaping, smart irrigation systems, and green infrastructure, such as swales, to manage stormwater.

When designing new buildings, they optimize energy through use of high-efficiency lighting, heating, cooling, and ventilation. They also carefully select windows, shading, and roofing materials to reduce heating and air-conditioning loads. They use LED lighting, with motion sensors to ensure lights are only being powered when required, and they use upgraded heating, ventilation, and air-conditioning systems to optimize energy efficiency. Their Prineville Data Center has a PUE of 1.09 [10].

Between 2011 and 2016, as a result of Facebook's combined work on data center efficiency and renewable energy procurement, they avoided more than 1.2 million metric tons of carbon dioxide equivalent (CO_2e); 862,000 metric tons were avoided as a result of more efficient operations, and 374,000 metric tons were avoided as a result of clean and renewable energy (CaRE) procurement [11]. That's the equivalent of taking over 256,000 passenger cars off the road for 1 year.

Challenge 2: One Company Alone Is Not Enough
Facebook works hard to drive efficiency and manage energy costs in its own operations and facilities. While they have an ambitious, science-driven goal of 100% renewable energy, they have found it difficult to purchase enough green energy to power their facilities and data centers in the USA and around the world.

Solution 2: Leading Collaborative Initiatives with Other Companies
Facebook decided to work with other large energy users (facilitated by nonprofit organizations) to organize collective purchasing power that is leading to new sources of renewable energy coming on line around the country and around the world. Facebook is a leader in organizing initiatives to make computing less expensive and less energy intensive, so it made sense to work with others to organize the purchasing power of many large energy users for renewable energy. In 2011, they helped found the Open Compute Project (OCP), a global community of engineers whose mission is to foster more openness, more innovation, and a greater focus on energy efficiency in computing technologies (Fig. 9.2). Today, there are more than 200 OCP members. Through OCP, they have shared the full physical layers of their data center stack—building, cooling, power, battery, open rack, networking, compute, and storage. By open sourcing their hyper-efficient designs, they enable companies of all sizes to use energy more efficiently and drive efficiency improvements across the industry. They believe the aggregate impact of OCP has the potential to provide significantly more leverage than the actions of any single company.

In their server designs, they eliminate any material that does not serve a specific function for the hardware. This eliminates nearly 6 pounds of material per server, which reduces the amount of materials needing to be produced, transported, assem-

Fig. 9.2 Seeking to transform the energy efficiency of global data centers, in April 2011, Facebook launched the Open Compute Project, an initiative to share the custom-engineered technology in its first dedicated data center in Prineville, Oregon. (Source: Facebook [12])

bled, and eventually disassembled. They also minimize the overall part count, accelerate assembly, and remove elements like front panels, paint, and logos.

> To truly make an impact on the world, we are committed to innovating fast, sharing our breakthroughs with others and collaborating to accelerate our collective progress.
> –Bill Weihl, Director of Sustainability, Facebook [13]

Challenge 3: Not Enough Renewable Energy on the Grid in the USA
In 2012, Facebook set a goal of reaching 25% clean and renewable energy (CaRE) in their electricity supply mix in 2015 for their operations. They exceeded this goal by reaching 35% in 2015, so they doubled that target and are now aiming to have at least 50% clean and renewable energy in their mix in 2018. Their long-term goal is to have 100% clean and renewable energy. This is challenging because they want solutions that result in new renewable energy brought onto the grid. Facebook directly purchases renewable electricity to reduce their footprint rather than simply buying offsets that may reduce emissions this year but won't have the long-term impact of helping build a clean energy market and future.

Solution 3A: Implement and Use Renewable Energy Tariffs
As of September 2017, there are renewable energy tariffs available in 13 states and the District of Columbia that help make renewables accessible [14]. Facebook played a role in the creation of several of those tariffs. In October 2017, Facebook announced another renewable energy tariff, which will enable them to power their Henrico Data Center in Virginia with 100% clean and renewable energy [15].

A key criterion in their site selection process for new data centers is access to clean and renewable energy. Their last nine data centers are (or will be once construction is complete) powered by 100% clean and renewable energy, including wind, solar, and hydro.

Solution 3B: Helping to Build the REBA Coalition
Founded in 2016 [16], the Renewable Energy Buyers Alliance (REBA) [17] is a coalition of nonprofit organizations—Business for Social Responsibility (BSR), Rocky Mountain Institute (RMI), World Resources Institute (WRI), and World Wildlife Fund (WWF)—that empower multinational corporations to transform electricity systems with renewable energy. REBA facilitates solutions among customers, renewable energy suppliers, utilities, and policy-makers to overcome market barriers and drive collaboration among all parties. Energy buyers like corporations, cities, and public institutions want to power their operations with clean energy, but navigating the path to renewables in the USA and international markets is harder than it should be.

Since the beginning, Facebook has been an active leader and driver in the initiatives that make up REBA. They have helped build the coalition of over 100 large buyers, including HP, Inc., Google, Intuit, Genentech, and VMware, to move markets and scale their impact and are committed to doing what they can by collaborating across sectors and geographies. Whether sharing what they are learning from bringing renewables to the grid or advising companies looking to green the grid, they make themselves available to members of REBA's initiatives. By helping to found REBA, Facebook is making strides in meeting their own renewable energy goals and also helping to transform provision of renewable energy in the USA and around the world.

Driving Force: Customer Collaboration to Transform Energy Markets
Facebook is instrumental in building alliances to advance more energy-efficient computing technology and data center design. Similarly, they are a pioneer in working with NGOs to bring together large energy users to share best practices and to advocate for better policy and regulatory practices. This is catalyzing the availability of more green electricity in more markets both around the USA and globally, as well as ensuring that the energy that Facebook uses is green.

9.3.4 *Google*[9]

"Google was thrilled to announce that in 2017 Google will reach 100% renewable energy for our global operations—including both our data centers and offices. This is a huge milestone. We were one of the first corporations to create large-scale, long-term contracts to buy

[9] This case study was developed in collaboration with staff from Google who provided information between Nov 2017 and March 2018 and approved the chapter. It incorporates material from white papers and other materials on the Google website, including Google's white paper from December 2016 "Achieving Our 100% Renewable Energy Purchasing Goal And Going Beyond" https://static.googleusercontent.com/media/www.google.com/en//green/pdf/achieving-100-renewable-energy-purchasing-goal.pdf Last visited 5/20/2018.

> renewable energy directly; we signed our first agreement to purchase all the electricity from a 114-megawatt wind farm in Iowa, in 2010 [18]. Today, we are the world's largest corporate buyer of renewable power, with commitments reaching 2.6 gigawatts (2,600 megawatts) of wind and solar energy. That's bigger than many large utilities…
>
> …The science tells us that tackling climate change is an urgent global priority. We believe the private sector, in partnership with policy leaders, must take bold steps and that we can do so in a way that leads to growth and opportunity. And we have a responsibility to do so — to our users and the environment…." Urs Hölzle, Senior Vice President of Technical Infrastructure, Google [19]

Google is one of the world's biggest electricity users. They are working hard to minimize their negative environmental impact and to control the cost of energy to their operations. This case study highlights the extensive and innovative approach Google has taken to bring more renewable energy online and to encourage the electricity industry to become more customer-focused, nimble, and responsive. They share their best practices with others through publishing white papers and detailed information on their website about their energy practices.

In 2015, Google consumed 5.7 terawatt hours (TWh) of electricity across all their operations [20], which is nearly as much electricity as San Francisco used in the same year [21]. Google's data centers—the engines of the Internet that power all of their products and services—require a reliable and constant supply of electricity to ensure that Google can deliver its many services, such as Search, Gmail, YouTube, and Maps, wherever customers need them. Electricity is one of the largest operating expenses at Google's data centers. Ensuring a cost-competitive, predictably priced electricity supply is an important business objective. The greenhouse gas (GHG) emissions associated with the electricity Google uses for their operations represent the majority of their total carbon footprint.

Challenge 1: Science-Driven Goals to Combat Climate Change
The science of climate change has shown that achieving absolute reductions in annual GHG emissions and sustainable long-term levels of GHGs in the atmosphere is an urgent global imperative. Businesses have an important role to play in driving robust, sustained action to transition to a clean energy economy.

Solution 1A: Power the World's Largest Energy User with 100% Renewable Energy
Google is committed to being part of the solution to solving global climate change, both through purchasing renewable energy to match the energy use of their own operations and by helping to create pathways for others to purchase clean energy themselves. Purchasing energy from renewable resources makes good business sense because they are cost-effective, and the inputs like wind and sunlight are essentially free. The cost to produce renewable energy technologies such as wind and solar has come down precipitously in recent years. Since 2010, the levelized cost of wind has come down 60%, and the levelized cost of solar has come down 80% [22]. In fact, in some areas, renewable energy is the cheapest form of energy available on the grid [23].

Buying energy from renewables gives Google great financial certainty. Having no fuel input for most renewables allows Google to eliminate their exposure to fuel price volatility and to smooth their financial planning over the long term. This means they can enter into long-term fixed-price agreements that set the price they pay for renewable energy from the start of a contract, avoiding volatility in their energy expenditures. This is especially helpful when managing a global portfolio of operations in a wide variety of markets, including 70 offices in 50 countries, as well as numerous data centers.

In 2012, Google made a commitment to purchase enough renewable energy to match 100% of their operations, and they achieved that goal just 5 years later in 2017. Google defines its 100% renewable purchasing goal as buying on an annual basis the same amount of MWh of renewable energy—both the physical energy and its corresponding RECs—as the amount of MWh of electricity that Google consumes for their operations around the world.

Solution 1B: Direct Purchases and Utility Partnerships for New Renewable Energy

Google makes direct purchases from renewable developers and forms partnerships with their utility providers to purchase renewable energy. Google is the largest corporate renewable energy buyer on the planet, directly purchasing 2.6 gigawatts (GW) of renewable energy in 2016. They have signed renewable energy contracts in five countries across three continents. These contracts have driven the construction of renewable energy projects and will generate more than $3.5 billion in capital investment by renewable energy project developers. The projects will also provide tens of millions of dollars in local taxes for communities and lease revenues for landowners.

Challenge 2: Power Operations on Region-Specific 24/7 Basis with Clean Energy

In addition to continuing to move forward aggressively with renewable energy technologies like wind and solar (Fig. 9.3), Google intends to achieve the much more challenging long-term goal of powering their operations on a region-specific, 24/7 basis with clean, zero-carbon energy. This will drive clean energy from being an important but limited element of the global electricity supply portfolio today to a resource that fully and completely powers both their operations and the entire electric grid of the future.

Solution 2A: Regional Renewable Energy and Energy Services: Purchasing and Policy

As Google continues to pursue renewable energy purchases as they grow, they will focus even more on regional renewable energy purchases in the local markets where they have data centers and operations. They will broaden the scope of energy sources to include technologies or services that enable 24/7 clean energy. They will also work to promote policies that empower energy consumers to choose their energy supply, which Google believes will help accelerate the transition to a 100% clean electricity grid while also driving economic growth.

Google has been carbon neutral since 2007, when they started purchasing robust carbon offsets to temporarily "neutralize" emissions from their operations. As their

Fig. 9.3 Google co-owns the 377 MW concentrated solar thermal Ivanpah facility in California. (Image: Google. Used by permission)

renewable energy PPAs have come online in the past 7 years, they have used this newly created clean energy to reduce their environmental footprint. This has enabled them to purchase proportionately fewer carbon offsets as their renewable energy purchasing has scaled up.

Google's 100% renewable energy purchasing goal means that they buy the same MWh of physical renewable energy—and its corresponding RECs—as the total amount of MWh of energy that they consume for their operations around the world. They measure performance to the goal on both annual and global bases. Annually, they compare renewable energy supply with consumption for their operations, summing all the purchasing and usage over a calendar year, so their total global renewable energy purchases match up 100% of their global consumption. They don't yet have a renewable supply source on each grid where they have operations, though this is their goal. Google's blog post [24] describes in detail what it means to reach their 100% goal, as well as why they don't build renewable energy projects on their data center sites and how they use the grid to manage the intermittency of renewables.

Solution 2B: Achieving Google's 100% Renewable Energy Purchasing Goal
Google uses four main purchasing tactics to supply their operations with 24/7 clean energy:

- *"Direct" renewable purchasing*: In Europe, deregulated wholesale and retail power markets make it possible for Google to directly purchase renewable energy and have it delivered to their data center retail bill using the local grid. They sign a PPA with a project developer on a grid where they operate a data center, as well as a separate "balancing agreement" with a competitive power market entity that helps deliver the PPA across the grid and that can also "firm and shape"[10] the energy so that they have constant, 24/7 electricity.

- *"Offsetting" renewable PPAs (aka "fixed-floating swaps")*: In geographies with regulated retail markets but deregulated wholesale markets, Google purchases renewable energy at the wholesale level, retires the associated RECs, and sells the power back into the same grid from which they later draw power at the retail level. They describe this structure in their white paper about Google's green PPAs [25].
- *Utility renewable energy tariffs*:[11] In areas where retail markets are not open to competitive suppliers and particularly where there is no auction-based wholesale market, Google worked with their utility provider to create a new class of rates called a "renewable energy tariff," in which the utility procures renewable energy on their behalf (described in their renewable energy tariff white paper) [26] for sale and delivery to Google.
- *Grid-mix renewable content*: In most regions where Google has consumption, the utility's grid mix contains energy from renewable resources that is not otherwise being purchased by specific consumers but rather is part of the "residual" mix. For each MWh of retail electricity that Google consumes at their data center, they count the portion that comes from residual renewables on the grid toward their 100% goal.[12] Further, Google continues to support efforts to reduce the carbon emissions of the utility generation mix where they operate.[13]

When pursuing any of the first three approaches, Google either leads or actively collaborates with a utility in the procurement process. This allows them to apply very high standards for what types of renewable energy projects they count toward their 100% goal.

Solution 2C: Google Applies Three Key Criteria in Selecting Renewable Energy Projects

- *Additionality*: To ensure that Google is the driver for bringing new clean energy onto the grid, they insist that all projects be "additional." They seek to purchase

[10] "Firming and shaping" is the process of combining the output of a non-dispatchable, intermittent resource (like a wind or solar farm) with the output of a dispatchable, non-intermittent resource (like a traditional gas-fired facility) over time to create a flat, constant electricity supply profile for a consumer. Retail electricity providers do this by purchasing enough firming energy on a wholesale market in a given hour to "balance out" whatever energy is being produced by a renewable resource, thereby ensuring that their customer is receiving a flat supply of electricity.

[11] A "tariff" is a utility industry term for a rate schedule.

[12] Data on the renewable content in a given utility's grid mix is provided to Google by the Emissions & Generation Resource Integrated Database (eGRID), which is managed by the US Environmental Protection Agency.

[13] For example, Google joined with Amazon, Apple, and Microsoft to file a legal brief supporting the Environmental Protection Agency's 2014 Clean Power Plan, which aims to accelerate the transition to cleaner sources of electricity. More information is here: https://blog.google/topics/environment/google-unites-with-other-tech-companies/. Last visited 5/20/2018.

energy from not yet constructed generation facilities that will be built above and beyond what is required by existing energy regulations (like state renewable energy standards). This approach also helps advance new technologies and drive economic growth in the regions where they operate.

- *Physical energy bundled with its "renewable certification"*: Google purchases both the physical electricity and its corresponding "bundled" certification of renewable energy—a renewable energy certificate (REC) in the USA or a guarantee of origin (GoO) in Europe.[14] By purchasing physical energy bundled with these certifications, Google provides all or nearly all of a project's cash flow over time, whereas buying "unbundled" RECs or GoOs provides only a small portion of a project's cash flow. They take care never to buy "unbundled" or "naked" RECs or GoOs, in which a renewable attribute is sold on an open market, independently of underlying physical energy.
- *Proximity*: Google seeks renewable energy projects that will operate on the same grids as their data centers. This creates a stronger physical and financial link between the renewable power they purchase and their operational electricity consumption.

Solution 2D: Wholesale Power Purchase Agreements—Google's First Renewable Purchasing Structure

Since 2010, Google has completed 20 purchasing deals totaling 2.6 GW of renewable energy. Pioneering this buying method at scale has been key to their ability to reach 100% renewable energy purchasing for their operations. Some of these PPAs are "direct" purchases in competitive retail markets (e.g., Europe and Chile) that allow Google to pass the obligations of a PPA they negotiate through to their retail bill at their facilities. It gives them the greatest degree of control over their energy supply contract and directly delivers the financial value of renewables to their bill.

Challenge 3: Fixed-Floating Swaps Not Scalable for Smaller Electricity Users

In areas with restrictive retail regulation, Google uses a fixed-floating swap approach. In these markets, they purchase power directly from a renewable developer at a long-term fixed price, retire the associated RECs, and then resell the physical electricity back into a wholesale market. Because they are locked into a single retail monopoly electricity supplier, they still must separately buy power from their local utility provider through a standard retail contract. Under the current regulatory model in most regions where they have significant operations, it is the most scalable way to rapidly procure renewable energy.

For all the benefits of fixed-floating swaps, however, the model also creates unnecessary layers of complexity and dilutes the financial benefits that they receive as an end user. Because of restrictive retail market structures, they are essentially

[14] In markets where renewable energy certification systems don't currently exist, Google is supporting their development. For example, Google provided a seed grant to the Center for Resource Solutions to begin laying the groundwork to establish REC systems in Asia, starting in Taiwan. More information is here: https://blog.google/topics/environment/laying-foundation-for-renewable-energy/. Last visited 5/20/2018.

buying power twice and selling it once—buying once at the competitive wholesale level and again at the regulated retail level, while they also sell at the competitive wholesale level. Since these two prices aren't always correlated, they don't reduce their exposure to market price volatility quite as much. Further, these structures also require significant resources and expertise to execute, as well as a long-term commitment from the buyer, so they aren't scalable options for many smaller companies that want to purchase renewable power.

Solution 3A: Utility Renewable Energy Tariffs: A First Step Toward an Opt-In Model for Purchasing Renewables

In an ideal world, large energy consumers like Google could simply ask their retail energy provider (whether competitive or regulated) to sell them the renewable energy they would like to buy. This means selling them competitively priced, scalable renewable energy from a newly constructed wind or solar farm within the time frame they need to meet the growing demand for data center services and with enough contractual flexibility to meet the needs of their business. It also means "firming and shaping" the renewables with dispatchable resources to deliver constant, 24/7 power.

To begin moving toward this model, Google has worked with their regulated utilities on programs to empower customers with options to purchase renewable energy. In 2013, Google worked with Duke Energy in North Carolina to create one of the first renewable energy tariffs, which allow customers in fully regulated territories to apply to their utility to purchase renewable energy from dedicated projects. Google is also working with stakeholders to create similar programs in Georgia and Taiwan.

Programs like these are first steps toward ensuring the electricity service model is agile and responsive to customer demands. Customers need only apply to their local utility provider to participate, making renewable energy accessible to a wide variety of customers—both large and small—without the need for these customers to have their own energy procurement teams. Further, because customers of these programs pay for the renewable energy from their retail bill, tariffs can be structured to more directly deliver the financial benefits that renewables provide.

Solution 3B: Making Green Tariffs Responsive to Customer Needs

Since 2013, Google has seen similar renewable energy tariff programs emerge in 13 states [14]. However, uptake of these programs is slow, and they remain largely unsubscribed. Many of these programs lack the responsiveness and agility that customers require; it can take many years from application to receipt of renewable energy supply. Moreover, the financial structure that customers need to make the business case to their management—cost-competitive renewables delivered directly and without unnecessary fees—is often lacking. Although these programs have been a helpful first step toward a purchasing structure that mirrors Google's ideal model, there is more work to do to make these programs as customer-centric as modern consumers demand.

Challenge 4: Utility Regulatory Regimes Do Not Provide Incentives for Utilities to Innovate
It's certainly possible for utilities in regulated markets to mimic the responsiveness and agility of competitive markets, but Google's experience is that this is difficult to implement. Most utility regulatory regimes do not provide the right incentives for utilities to meet customer calls for new products or to create innovative solutions to unique customer needs.

Solution 4A: Competitive Energy Markets Provide the Fastest Path to Purchasing Renewables
The flexibility, accessibility, and responsiveness to customer demands that come with competitive markets make it easy for customers to purchase the products they desire. Access to renewables for retail electricity customers is critical to the business case for renewable energy. Wholesale renewable energy PPAs have proved to be a successful method for large companies to rapidly procure renewables, but it is only a partial solution. Regulatory models must evolve to enable customers to directly place the benefits and costs of renewables—energy prices, capacity credits, etc.—on their retail bills.

Solution 4B: Utilities Need to Become Truly Customer-Centric to Meet the Demands of the Market
Renewable energy PPAs have worked for large customers like Google, but they are complicated. Without robust buy-as-you-go renewable energy programs offered by utilities and competitive energy suppliers, most consumers won't be able to opt in to purchase green energy. The lack of such options available in the market today is largely the result of a fundamental disconnect between the demands of modern electricity consumers and the regulated utility business models that were designed for a bygone era. Meeting the demands of today's market will require an evolution of business and regulatory models to align utilities' incentives so that they are rewarded for the value that they deliver to their customers.

Solution 4B: Cost-Competitiveness Is Critical to Unlocking Renewable Energy for All
In some geographies, renewables are now the cheapest source of energy available, with generation-weighted average levelized wind PPA prices that are below the average cost of wholesale energy on the grid. For example, average levelized wind PPA prices in 2014 in "interior" states (Arizona, Colorado, Iowa, Kansas, Minnesota, Missouri, Montana, Nebraska, North Dakota, Oklahoma, South Dakota, Texas, and Wyoming) were comparable to or lower than the average annual wholesale power prices in those states [23]. This precipitous drop in cost has been instrumental in shaping the business case for corporate renewable energy purchasing, and it will be key to the future success of the renewables sector.

Solution 4C: High Levels of Intermittent Renewables Can Be Managed Reliably and Cost-Effectively, But It Requires Larger, More Integrated, More Liquid Markets

There's much evidence to show that, despite their intermittent production, renewable energy resources like wind and solar can indeed be integrated into a grid at scale and at a low cost. In many US states, integrating intermittent wind resources into the grid often costs below $5 per MWh (a fraction of the overall energy cost), even for states with wind penetrations up to or exceeding 40% of peak load [27]. The Southwest Power Pool conducted a study in 2016 showing that it can reliably manage wind penetration levels of up to 60% with the addition of transmission upgrades and system tools [28].

The key to managing this intermittency at low cost has been the ability to use large, interconnected, highly integrated electricity grids and associated liquid wholesale markets. As renewable penetrations grow, it will be critical to shift from balkanized, isolated electricity markets to regional, interconnected grids and markets. This will create larger balancing areas to better manage intermittency, increase price efficiency through greater liquidity and market transparency, and allow renewables to be delivered from distant but resource-rich geographies to the load centers where they are needed.

Challenge 5: Achieving a Truly Zero-Carbon Electricity Grid Requires Addressing Variability

For Google to ultimately tackle the emissions associated with their electricity consumption, they need to move beyond their global, annual matching method to ensure that hour by hour, their operations are powered by clean energy on a region-specific basis. A key hurdle remains the variability of renewable energy technologies like wind and solar.

Solution 5A: Take a Regional Approach to Renewable Energy Procurement

Today, Google overbuys renewable energy in regions like the central United States and Northern Europe, taking advantage of favorable market conditions to purchase enough renewable energy to offset their inability to purchase it in other locations. In order to meet the longer-term challenge of matching their electricity consumption with clean energy, they will shift efforts more toward purchasing renewables and other forms of clean energy on the same grids where they operate.

Google is actively looking for renewable energy projects in every grid where they have significant operations. In many of these markets, there is currently no path to purchase renewables, so they will work with utilities and other local stakeholders to advance new purchasing pathways, as they have done in North Carolina and elsewhere. Importantly, on each grid, they will strive to have a direct financial tie between specific clean energy resources and their final retail energy bill, as they are able to do in places like Northern Europe and Chile. This is critical to the business case for renewable purchasing.

Solution 5B: Pursuing Technology Approaches that Advance 24/7 Clean Energy

To achieve Google's ultimate goal of 24/7 clean energy, they will need to address the variability associated with renewables in their energy supply portfolio through advanced technological approaches, among other means. In the future, they may pursue dispatchable, zero-carbon generation energy options for their portfolio, as well as direct purchase of wind and solar electricity. This could include purchasing energy from technologies like renewables paired with utility-scale energy storage, advanced nuclear power, geothermal energy, low-impact hydro, demand response and energy efficiency resources, or others. Google expects that a combination of technical innovations, economies of scale, and public policy will speed the pace of adoption and bring costs down over time.

Solution 5C: The Role of Government—Promoting Energy Policies that Empower Consumers and Accelerate the Transition to 24/7 Clean Energy While Driving Economic Growth

Google recognizes that establishing an advanced energy and climate policy architecture is critical to driving electricity grids around the world to become 100% clean, resilient, customer-centric energy systems. Through a variety of advanced energy policies, governments can help accelerate the transition to a clean energy grid. They can help the market correctly price the full array of impacts and benefits from energy technologies—particularly the harm associated with GHG emissions. Governments should help markets account for the impacts of these emissions—for example, by creating revenue-neutral, economy-wide mechanisms that "internalize the externality" of GHG emissions. Governments can also scale up investment in research and development to drive down the cost curves for emergent but expensive technologies while also ensuring that we don't miss the next breakthrough energy technology. Financial incentives that facilitate deployment can help bring energy technologies to maturity and generate economic growth. Investments in energy infrastructure like transmission lines can help bring clean energy from resource-rich geographies to the load centers where it is used.

Lastly, making retail and wholesale energy markets more nimble and customer-centric is critical to enabling any electricity customer who wants clean power to purchase it. This means evolving utility regulation and business models at the state and regional levels. Google's experience has shown that energy markets that prioritize customer choice at the retail level and unlock cost-efficiencies through large, regional markets at the wholesale level are effective at rapidly scaling up clean energy while delivering many other benefits to consumers.

Driving Force: Policy and Regulatory Change to Make Renewables Readily Available

Once Google committed to 100% GHG-free electricity, they worked with energy developers, utilities, and regulators across the USA and around the world to be able to purchase renewable energy. This proved to be much harder than they expected. Policy and regulatory changes are still needed to allow retail customers to be able to purchase green electricity in all markets.

9.3.5 *Stanford University*[15]

> "If you can find a cheaper better way to do things, why not do it! There is hope that projects can move fast with proper support."—Joe Stagner, Executive Director of Sustainability and Energy Management for Stanford University"[16]

In 2005, Stanford University began becoming more aware of sustainability. Rather than asking each campus department to take responsibility separately for their own environmental footprint, the University created a sustainability department to create an energy and climate plan to address the use of energy, water, waste, and other environmental factors across the entire institution. This made it easier to tackle shared services such as energy, landscaping, building design, and operations. Joe Stagner joined as Executive Director of Sustainability and Energy Management in 2007. He was tasked with developing options for the campus to provide energy and water in a more sustainable way. They found that the biggest challenge by cost and environmental impact was the natural gas-powered cogeneration plant responsible for 95% of the University's GHG emissions. The remaining GHG emissions were from sources such as the vehicle fleet, lab gases, and stationary generators. In parallel with energy, they worked to reduce use of water across the campus (Fig. 9.4).

Challenge 1: Institutional Barriers to Change
The sustainability team set out to overcome the institutional barriers that often get in the way of real change. Decision-makers would want to know that the team had evaluated a wide range of options and that they had sound engineering and economic analysis to support their plan. They had to show a strong business case and work with financial officers to determine how best to finance the project. The team worked to evaluate their current system; consider a wide range of options; model possible solutions, systems performance, and costs; and work hard to engage a range of stakeholders and decision-makers throughout the process.

Solution 1: Adopting a Long-Term Planning Horizon
University management decided to take a long-term planning horizon of 2015–2050 to address energy demand management and supply options because infrastructure would last for 30 or years or more. Third-party Cardinal Cogeneration owned the cogeneration plant whose contract was due to expire in 2015. The team knew it would take 5–6 years to plan replacement of the old plant. This also allowed them to make their planning horizon match the state and international targets to reduce GHG emissions by 80%, set for 2050.

Challenge 2: Need for Detailed Energy Use Data
The next challenge was to fully understand the hourly use of energy by the campus, 24 h per day, 365 days per year. Stanford wanted to build an economic model to test various solutions on the energy supply and demand sides. Which solutions and

---

[15] Case study based on a personal interview with Joe Stagner, Executive Director of Sustainability and Energy Management for Stanford University, conducted on November 27, 2017, at the Stanford Central Energy Facility, Stanford, California, and materials from the Stanford Energy website.

[16] Personal Communication. November 2017.

Fig. 9.4 Entrance to Stanford Central Energy Facility, Stanford University, California. (Stanford photo from website used by permission)

investments would be most cost-effective? How much energy are they generating and how is it being used on an hourly basis? If you are going to use thermal storage at night and deliver heat during the day, you need to know the daily cycle of demand by hour over 24 h in a day.

Solution 2: Acquire and Analyze Detailed Energy Data (24/7/365)
At first the energy provider did not want to provide such detailed information; it took nearly 6 months to get hourly data manually from the central plant. Once Stanford had the data, they discovered a huge heat recovery opportunity, which revealed options they didn't know existed. This underscored that detailed information helps make better decisions.

Challenge 3: Modeling Costs Accurately
Getting the business model right was critical. Stanford has a longtime horizon, so they conduct life cycle assessments of capital investments rather than imposing an arbitrary payback period. They worked with a cost of capital 5% discount rate and 2.5% inflation. The University can borrow money through bonds at 4.5% paid back through operations savings. Reimbursement for research covers one-third of the cost of running the University so they needed to recover true cost of utilities for delivering services.

Solution 3A: Separate Cost of Utility Infrastructure from Cost of Buildings
The total cost of getting energy to the building includes infrastructure and debt service. It was important not to lump debt for energy with buildings to avoid making bad long-term business decisions. Even with a good financial model, some institutions have debt limitations for borrowing capacity based on building stock. Separating utility infrastructure from buildings enables separate financing from the cost structure of the buildings. Capital infrastructure can be repaid through utility rates. The institution can source separate revenue bonds, which can be paid back through utility billing, financing physical improvement to the system with a lower long-term debt structure.

Solution 3B: Build an Economic Model for Campus-Wide Energy Use
Stagner and his team built an overall economic model for energy use on the campus, looking at electricity, heating, and cooling demand and how much it cost. They considered how energy efficiency improvements would reduce the demand for heating and cooling and how the energy would be supplied to meet different loads. They considered the impact of reducing demand on GHG emissions and energy costs. The model allowed them to also hold demand constant and model different types of energy supply and to experiment with a variety of potential investments on both the demand and supply sides. They assumed long-term growth of demand on the campus, using the known campus capital plan and general use permit growth rates. They developed heating, cooling, and power load forecasts and tested them under various scenarios, including a business as usual case.

Challenge 4: Trade-Offs Between Efficiency, Retrofits, New Buildings, and Green Sourcing
Stagner's team considered what level of investment in energy efficiency design to specify in new building construction to evaluate what the cost premium would be against the incremental load of less efficient buildings. What would be the best value for investment in buildings: 30%, 40%, or 50% more efficient than ASHRE 90.1 code?

Solution 4: Assess Portfolio for Tradeoffs Between Efficient New Design and Retrofit Projects
The team's analysis showed that 30% below (more rigorous than) energy code was most cost-effective for new building design. Otherwise, retrofits of existing buildings or energy supply changes would be more cost-effective than building the most energy-efficient new buildings for their overall portfolio.

Challenge 5: Stakeholder, Expert, and Decision-Maker Engagement
During the planning process, the team reached out to faculty, students, staff, outside experts, and trustees to solicit ideas and to get feedback and to build buy-in, especially from University leaders. They looked at all types of possibilities, including wave power, wind, solar, and more. One idea they explored was heat pumps and heat recovery. Traditionally, buildings are heated in winter and cooled in summer. If there is a thermal overlap, you may be able to use energy to cool or heat at alternate times. That potential opportunity drove the request for hourly energy demand data. They expected a 5–15% overlap in total heating and cooling to prevent wasting heat. But the data showed that there was a 75% opportunity for heat exchange on a daily basis. One system was providing heat, and the other system was throwing it away (providing cooling and dehumidification). They asked themselves if an energy system could be devised to use waste heat to meet the heating loads for the campus? Managing both systems together became the key to a new, more sustainable, and economical approach.

Solution 5: Making the Case and Getting Buy-In Ahead of Time
When the time came to recommend a bold solution that required a major capital investment, senior management support was already on board. The team had

recruited respected advisors, taken the time, collected the data, and modeled options to educate decision-makers. Stagner showed that if you consider what makes sense both economically and sustainably, you can be empowered to do things that are better and cheaper over the life of the infrastructure.

Stanford selected the most economic and most sustainable outcome, allowing it to pollute less and save money. They determined that the University had the borrowing capacity to finance a new energy system for nearly $500 million without restricting the University's ability to meet other important needs. The capital cost of $485 million will save more than $420 million over 35 years.

This thorough process of analysis, review of options, and innovative thinking led to the Stanford Energy System Innovations (SESI). SESI has transformed Stanford's energy supply from a 100% fossil fuel-based combined heat and power plant to grid-sourced green electricity and a more efficient electric heat recovery system. SESI includes high-efficiency standards for new buildings, continued efficiency improvements for existing buildings, and the cutting-edge energy supply system.

Challenge 6: Discharging Heat Through Cooling Tower Wastes Energy
The Stanford campus supports a range of pursuits from research and dormitories to athletic facilities. This results in the campus having a 70% real-time overlap of heating and cooling demands and presents the opportunity for heat recovery—using waste heat collected by the chilled water system to meet the University's concurrent heating needs.

Solution 6: Capture and Reuse "Waste" Heat Through Heat Recovery
The new heat recovery system collects waste heat from buildings via a chilled water loop and captures it at the central energy facility (CEF) for reuse, eliminating the use of cooling towers to discharge the heat. Instead, heat recovery chillers move waste heat collected from the chilled water loop to a new hot water loop that distributes heat to the buildings. The heat recovery system meets 88% of the heating load on campus with waste heat and reduces total campus water consumption by 18% (Fig. 9.5).

While this system works well in the mild climate of Silicon Valley, further analysis shows that electrification using renewable energy and heat exchange systems could be a globally scalable solution. For example, Stanford researchers determined that for the University of Illinois, 55% of heating could be matched with heat recovery and for a major research university in New England, 49% heating could be matched with their own heat recovery, even though those institutions are located in regions with greater weather variability and temperature extremes.

Stanford University pursues innovation in energy and sustainability in the lab and in daily operations. Their goal is to power the University with 100% renewable energy. Additional measures to reduce GHG emissions include electrifying its fleet of buses and encouraging students, faculty, and staff to use alternative forms of transportation to single-occupancy cars. They provide detailed information on SESI and other sustainability programs to share what they are learning and doing to encourage widespread adoption. As a result of these efforts, Stanford is now one of the most energy-efficient research universities in the world.

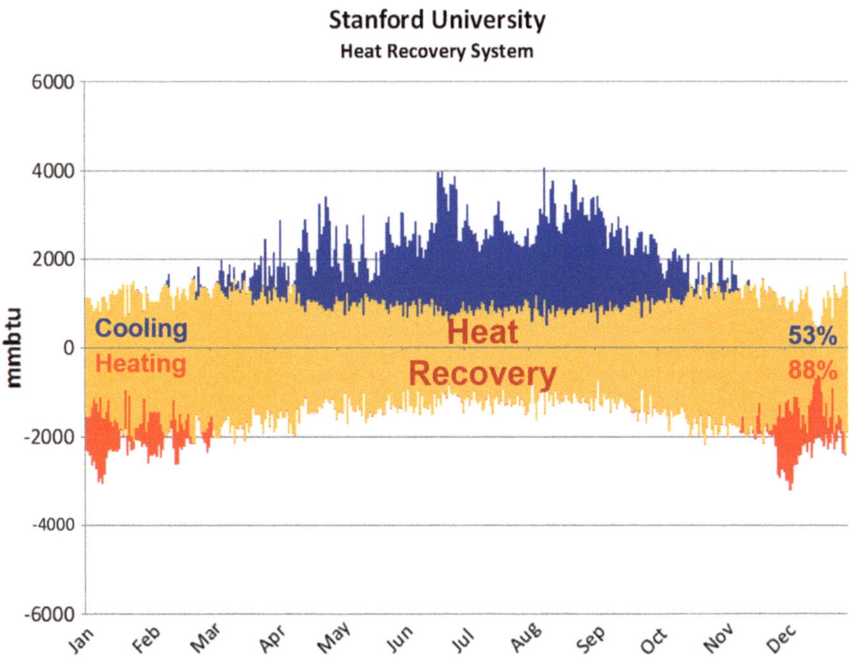

Fig. 9.5 Visualization of Stanford's heating and cooling demands [29]

Driving Force: Detailed Energy Data, Modeling, and Leadership Engagement
By looking at very detailed energy use for heating and cooling by each building over a year, the Stanford energy team was able to discover a substantial overlap and to radically rethink how to deliver comfort and other energy services to building occupants while dramatically reducing GHG emissions and operating costs. They engaged University leaders throughout the process so that approval of a major capital investment could be obtained and the project completed in record time. A holistic and practical approach led to breakthrough results.

9.4 Summary

This chapter explored five case studies about the challenges Silicon Valley electricity customers faced and potential solutions for implementing renewable energy (Table 9.1).

Climate change is a pressing challenge for humanity. It is a problem that can only be solved through a transformation of our energy systems. New technology is critical to enabling this transition. All of the large Silicon Valley institutional customers in this chapter, whether corporate or academic, overcame obstacles to pursue and achieve their goals of GHG reduction through clean electricity. They set goals, worked with utilities and regulators, improved their own operations, implemented

Table 9.1 Silicon Valley challenges and solutions for implementing renewable energy

Challenges for Silicon Valley Electricity Customers	Solutions for Silicon Valley Electricity Customers
The utility industry is heavily regulated, which can make responding to customer demand difficult and slow	Setting science-based goals for reducing or eliminating GHG emissions
Old beliefs about the cost of renewables and energy efficiency are out of date	Leadership considers sustainability and good global citizenship to be important
Utilities typically provide limited data to customers. Customers may not fully understand their energy use. Many opportunities are missed due to lack of detailed data	Hourly energy consumption data is necessary to design an efficient energy system. This takes smart technology. Detailed data is the foundation for innovative solutions
Corporate real estate often relies on leases lasting 3–5 years and may occupy space in multi-tenant buildings. This inhibits direct collaboration with utilities to reduce energy use and to advocate for greener electricity	Green leases can align the interests of tenants, property managers, and property owners. Utilities can work with large energy users to provide energy usage data from smart meters and other sources to help prioritize improvements and to reduce demand
Accounting systems that require paybacks sooner than the life of the capital infrastructure	Robust business case that takes life cycle costs into account and separates building finance from infrastructure finance for campus-based systems
Decision-makers may be operating on out-of-date assumptions and lack of detailed knowledge and analysis	Involve key stakeholders and decision-makers in understanding analysis and in building the business case for implementing transformational new energy systems
The utility tariff structure is inflexible and does not facilitate bringing new renewable energy onto the grid. Regulation inhibits innovation and responsiveness	Work with multiple stakeholders to share what they are learning and to bring collective market power to the utilities and to the regulators
Systems such as heating and cooling, building design, and other infrastructure are managed separately	Holistic approach to green building design and upgrades is more effective than a siloed approach. Many measures have multiple benefits
Natural gas infrastructure is in place, and natural gas is relatively inexpensive	Electrifying heating and cooling using heat exchange creates less carbon and is less expensive than natural gas
Large carbon footprint from vehicles with internal combustion engines. Transition to EVs requires investment in vehicles and charging infrastructure	Electrifying transportation, powered by 100% carbon-free electricity lowers operating costs. Institutions provide EV charging stations for employees
Many utilities see their business as managing load and not as meeting customer needs for advanced energy management, including detailed energy data, green energy sources, and dispatchable services	The future of electricity will be more like information, managed dynamically and flowing freely between a wide variety of sources, including energy storage and demand management, relying on partnerships between energy producers and users

new technology, and joined with others to combine market demand for new, clean renewable energy resources. Regulators and utilities are learning to deliver new levels of service responsive to customer needs. Along the way, many parties shared their experiences and breakthroughs with others to pave the way for true transformation of how energy is produced and used, in order to protect the long-term well-being of our communities and our shared planet.

Author Profile

Marianna Grossman is Founder and Managing Partner of Minerva Ventures, a consultancy focused on solutions for a resilient future. Minerva advises companies and communities on climate risk, adaptation, and resilience and guides clean-tech startups in water, energy, remediation, finance, and climate analytics. Marianna brings people and institutions together across sectors to collaborate, invent, and implement solutions that enhance sustainability and address the consequences of climate change. Previously, Marianna led Sustainable Silicon Valley (SSV) for 7 years. She worked in the automotive, computer, and semiconductor industries. Board service and membership includes California's Climate Adaptation Action Plan Technical Advisory Group, 2016; Sustainable Water Resources Roundtable; the international Balaton Group, Transportation Choices for Sustainable Communities; ClimateMusic Project; Sustainability Committee of the SF Bay Area Super Bowl 50 Host Committee; and climate action councils for the cities of Palo Alto and Mountain View. She earned a BA cum laude, with distinction in Policy Studies from Dartmouth and an MBA from Yale School of Management.

References

1. "VMware's Commitment to Sustainability: CEO Pat Gelsinger Message to Employees," June 2017. https://www.vmware.com/content/dam/digitalmarketing/vmware/en/pdf/company/vmware-commitment-sustainability-pat-employees.pdf. Accessed 20 May 2018.
2. " Green IT: Virtualization Delivers Energy and Carbon Emissions Reductions," IDC, October 2016, https://www.vmware.com/content/dam/digitalmarketing/vmware/en/pdf/sustainability/vmware-greenit-virtualization-delivers-energy-carbon-emissions.pdf. Accessed 20 May 2018.
3. "Data Center Energy and Carbon Emission Reductions Through Compute, Storage, and Networking Virtualization," September 2017, vmware.com/go/VMware-IDC-EmissionsReductions-Compute-Storage-and-Networking-2017. Accessed 20 May 2018.
4. Corporate Renewable Energy Buyers' Principles, http://buyersprinciples.org/. Accessed 20 May 2018.
5. RE100, http://there100.org/. Accessed 20 May 2018.
6. " VMware's CDP Climate Change Information Request," VMware, 2017, https://www.vmware.com/content/dam/digitalmarketing/vmware/en/pdf/sustainability/vmware-cdp-climate-change-information-request.pdf. Accessed 20 May 2018.
7. "Sustainability Report at HP," 2016 Sustainability Report, Introductory Letter from Dion Weisler, President and Chief Executive Officer, http://www8.hp.com/us/en/hp-information/global-citizenship/index.html#. Accessed 20 May 2018.
8. Facebook post by Mark Zuckerberg, 1 June, 2017, https://www.facebook.com/zuck/posts/10103765754210171. Accessed 20 May 2018.
9. http://www.opencompute.org/blog/water-efficiency-at-facebooks-prineville-data-center/ Posted 8/9/2012. Accessed 20 May 2018.
10. "Prineville, OR Data Center," Facebook, https://www.facebook.com/PrinevilleDataCenter/app/399244020173259/. Accessed 20 May 2018.

11. "Our Footprint," Facebook, https://sustainability.fb.com/en/our-footprint. Accessed 20 May 2018.
12. "Open Compute Project," Facebook, https://newsroom.fb.com/media-gallery/data-centers-2/server-room-dark1/. Accessed 20 May 2018.
13. "Building the most efficient data centers, on Earth," Facebook, https://sustainability.fb.com/data-centers/. Accessed 20 May 2018.
14. L. Tawney, P. Barua and C. Bonugli, *Emerging Green Tariffs In U.S. Regulated Electricity Markets*, World Resources Institute, September 2017. http://www.wri.org/sites/default/files/emerging-green-tariffs-in-us-regulated-electricity-markets-sep2017_0.pdf. Accessed 20 May 2018.
15. "Dominion Energy Virginia to Power New Facebook Data Center with Renewable Energy," Dominion Energy, 5 October 2017, https://www.prnewswire.com/news-releases/dominion-energy-virginia-to-power-new-facebook-data-center-with-renewable-energy-300531771.html. Accessed 18 May 2018.
16. " Renewable Energy Buyers Alliance Forms to Power the Corporate Movement to Renewable Energy," Renewable Energy Buyers Alliance press release, 12 May 2016, http://www.wri.org//news/2016/05/release-renewable-energy-buyers-alliance-forms-power-corporate-movement-renewable. Accessed 20 May 2018.
17. REBA, http://rebuyers.org/. Accessed 20 May 2018.
18. "Reducing our Carbon Footprint with the Direct Purchase of Renewable Energy," Google blog, 20 July 2010, https://googleblog.blogspot.in/2010/07/reducing-our-carbon-footprint-with.html. Accessed 20 May 2018.
19. "Our Values in Action," Google, https://www.google.com/intl/en/about/values-in-action/renewable/. Accessed 20 May 2018.
20. "Climate Change 2017," Alphabet Inc., https://www.cdp.net/en/responses?utf8=✓&queries%5Bname%5D=alphabet. Accessed 20 May 2018.
21. "Electricity Consumption by County," California Energy Commission, http://ecdms.energy.ca.gov/elecbycounty.aspx. Accessed 20 May 2018.
22. *Lazard's Levelized Cost of Energy Analysis Version 9.0*, Lazard, November 2015, https://www.lazard.com/media/2390/lazards-levelized-cost-of-energy-analysis-90.pdf. Accessed 20 May 2018.
23. *2014 Wind Technologies Market Report*, United States Department of Energy, August 2015, https://www.energy.gov/eere/wind/downloads/2014-wind-technologies-market-report. Accessed 20 May 2018.
24. "Understanding our goal: What it means to reach 100% renewable energy purchasing, Google blog, 8 February 2016, https://www.blog.google/topics/environment/google-green-blog-what-it-means-to-be_8/. Accessed 18 May 2018.
25. "Google's Green PPAs: What, How, and Why," Google, 17 September 2013, https://static.googleusercontent.com/external_content/untrusted_dlcp/www.google.com/en/us/green/pdfs/renewable-energy.pdf Last visited 5/18/2018. Accessed 20 May 2018.
26. *Expanding Renewable Energy Options for Companies Through Utility-Offered "Renewable Energy Tariffs,"* Google, 19 2013m https://static.googleusercontent.com/media/www.google.com/en//green/pdf/renewable-energy-options.pdf. Accessed 20 May 2018.
27. *2015 Wind Technologies Market Report*, United States Department of Energy, August 2016, https://www.energy.gov/eere/wind/downloads/2014-wind-technologies-market-report. Accessed 20 May 2018.
28. *2016 Wind Integration Study*, Southwest Power Pool, 5 January 2016, https://www.spp.org/documents/34200/2016%20wind%20integration%20study%20(wis)%20final.pdf . Accessed 20 May 2018.
29. "Innovation: Heat Recovery," Stanford University, https://sustainable.stanford.edu/sesi/innovation. Accessed 28 March 2018.

Chapter 10
Santee Cooper's Coal Ash Impoundment Closure Challenge: Risk Reduction through Recycling

Jay Hudson

Abstract After the Kingston ash spill in 2008, Santee Cooper, like many utilities, addressed closure of legacy coal ash impoundments. Grainger Station in Conway, South Carolina, was a 1960s vintage coal-fired facility with a legacy coal ash issue. Faced with mounting pressure to remove the ash from the impoundments, Santee Cooper developed an innovative solution to remove the ponded ash from the site and into recycling via cement manufacture. This solution involved engagement with both internal and external stakeholders while cleaning up groundwater, all completed at a similar cost to in situ closure.

10.1 Introduction

In 2013, Santee Cooper faced a problem decades in the making – how to safely close legacy coal ash impoundments at retired or soon to be retired coal-fired generating stations. The issue was complicated by the differing interests of the stakeholders involved, and it became a challenge to balance the needs of the local riverine ecosystem, regulatory agencies, customers, and the community. Could Santee Cooper create a unique solution that would be acceptable to all involved with minimal trade-offs?

When Santee Cooper was created in the 1930s, the Coastal Plain of South Carolina was for the most part rural. The 60-mile stretch of beaches from the South Carolina/North Carolina border to Georgetown known as the "Grand Strand" was just beginning to develop as a vacation destination with early resorts opening in the mid-1930s in and around what is now Myrtle Beach, South Carolina. The South Carolina Public Service Authority, also known as Santee Cooper, was created in the 1930s in part to serve these largely rural areas, including the Grand Strand.

When Santee Cooper first began to generate electricity in 1942, the only power source was a 130-megawatt (MW) hydroelectric facility located near Moncks

J. Hudson (✉)
Santee Cooper, Moncks Corner, South Carolina, USA
e-mail: jhudson2@comcast.net

J. Fox, M. Scott (eds.), *Sustainable Electricity II*,
https://doi.org/10.1007/978-3-319-95696-1_10

Corner, South Carolina, over 80 miles from Myrtle Beach. This provided enough generation for the rural areas Santee Cooper served until the late 1950s and early 1960s when growth, mostly in and around Myrtle Beach, required new generation. In 1966, Grainger Generating Station began producing coal-fired electricity. The 180-MW plant was located along the Waccamaw River within the city limits of Conway, SC, only a short drive from Myrtle Beach. Grainger Station was typical of coal plants constructed at the time, with once-through cooling and a large man-made cooling reservoir and wet coal ash stored in impoundments (Fig. 10.1).

The logical location for ash impoundments would be adjacent to the Waccamaw River and between the river and the plant structure. Adjacent ash impoundments, even those adjacent to a waterbody, would be typical industry best practice at the time. From 1966 to the 1970s and 1980s, Grainger Station supplied electricity to the growing Grand Strand and surrounding areas. The first ash impoundment was nearly filled by the mid-1970s, requiring construction of a second ash impoundment in 1977.

Fig. 10.1 Grainger Station showing coal pile and the original ash pond directly behind it

10.2 Coal Ash

Per the US Environmental Protection Agency (EPA), "coal combustion residuals," commonly known as coal ash, are created when coal is burned by power plants to produce electricity" [1]. Coal ash is separated into bottom ash, glassified heavier particles that fall into the bottom of the boiler and fly ash, which "flys" out of the boiler and is collected by an electrostatic precipitator or baghouse. The majority of ash generated is fly ash, and studies have shown while bottom ash is nonreactive and inert, fly ash has a tendency to leach heavy metals over time when stored wet [2]. Acidic waters from coal pile runoff are collected in the ash impoundments and contribute to metals solubility, making them more mobile and prone to transport to groundwater. The majority of the material stored in the ash impoundments was fly ash. The fly ash is released from the base of the precipitator or baghouse into a pipe where water is used to transport it to the impoundments by a process called wet sluicing. A typical coal plant waste stream layout is shown in Fig. 10.2.

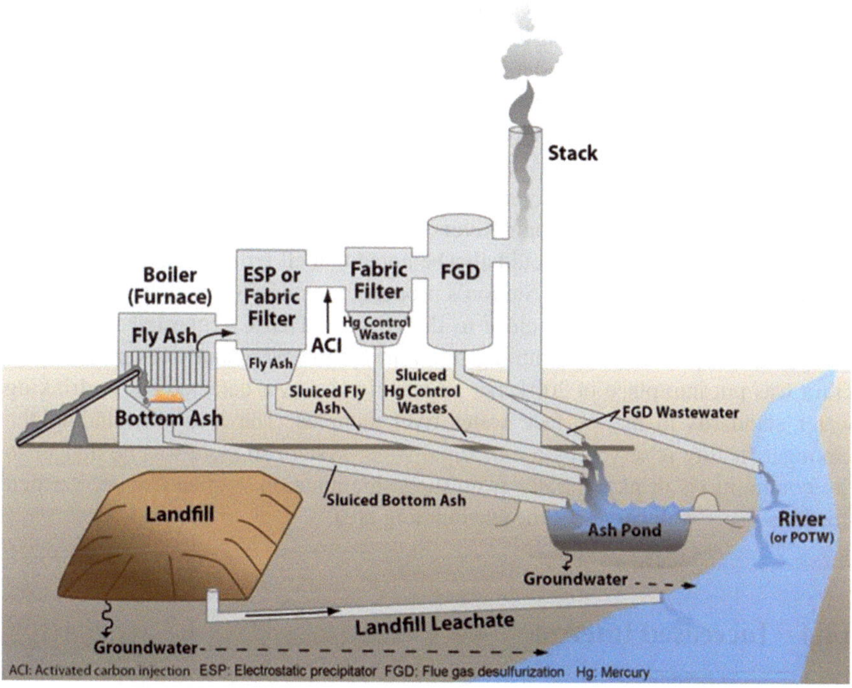

Fig. 10.2 Typical coal unit waste streams [3]

10.3 Groundwater Monitoring

As part of a Clean Water Act discharge permit renewal (referred to as National Pollution Discharge Elimination System or NPDES permits) in the mid-1990s, the South Carolina Department of Health and Environmental Control (SCDHEC) required installation of groundwater monitoring wells at various locations around the site to ensure that no polluted water was seeping into underground water flows (i.e., groundwater). Biannual groundwater monitoring was also required, and parameters included metals commonly found in coal ash, including arsenic. These installation and monitoring requirements were expanded to other coal ash impoundments throughout South Carolina as part of NPDES permit renewals throughout the 1990s and early 2000s.

South Carolina regulations require that all groundwater, including that contained in the superficial aquifer, meet drinking water standards. Up until 2001, the arsenic drinking water standard was 50 parts per billion (ppb). While groundwater from several of the newly installed wells was compliant with drinking water standards, groundwater from three others located between the ash impoundments and the Waccamaw River tested in excess of the drinking water standard for arsenic, which at the time was 50 ppb. Levels were as high as 10–16 times the drinking water standard in the initial testing. The metals contained within the fly ash along with the physical conditions within the impoundment were a suspected source.

Biannual groundwater samples continued to be taken at Grainger Station, with results reported to SCDHEC. Between 1996 and 2008, some groundwater samples tested in excess of 20 times the drinking water standard for arsenic. Any deviations above the drinking water standard were further exacerbated by EPA's lowering of the arsenic standard from 50 ppb to 10 ppb in 2001. During this time, SCDHEC and Santee Cooper held several discussions regarding the arsenic groundwater levels at Grainger Station. As mentioned above, it is interesting to note that many of the wells met drinking water standards, including wells close to the impoundments (but located away from the Waccamaw River), while wells closest to the river contained elevated levels of arsenic. It should also be noted that all groundwater impacts were contained within the Grainger Station's property boundary (Fig. 10.3). Because of the elevated levels of arsenic close to the river, SCDHEC requested that Santee Cooper develop an arsenic sampling program for the Waccamaw River. This program was put into place in 2004, but no arsenic was ever detected above drinking water standards in any surface water river samples. The area surrounding the Grainger facility is served by a municipal water system so there was no danger to the community's drinking water from the contamination. Santee Cooper verified this by a water well survey of the surrounding area.

10.4 Increased Interest

The devastating Tennessee Valley Authority (TVA) ash spill in Kingston, Tennessee, turned community and regulator focus on ash impoundments across the country, and Grainger was no exception (Fig. 10.4).

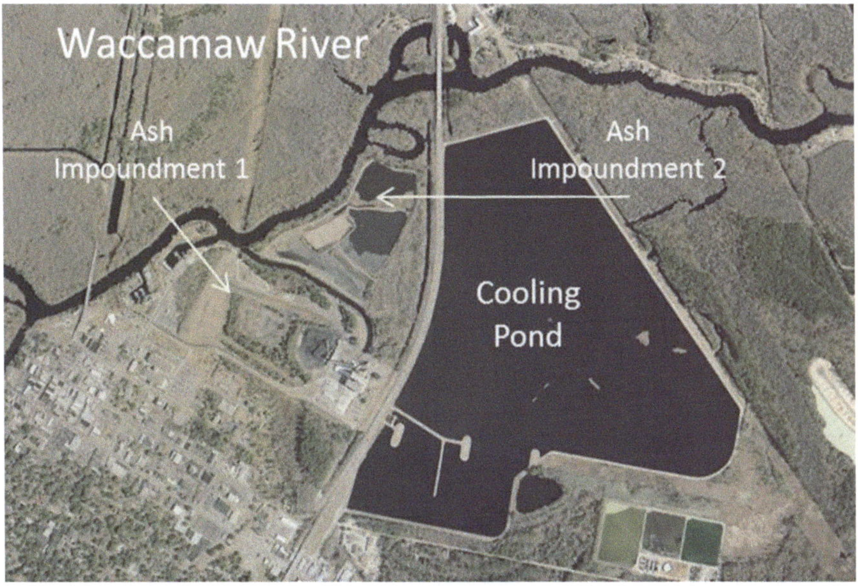

Fig. 10.3 Aerial view of Grainger Station

In early 2009, SCDHEC began regulatory actions to require groundwater mitigation and clean up at Grainger Station along with other coal ash impoundments at various coal-fired generating stations across South Carolina. In response, Santee Cooper employed consultants to study the issue and prepare a course of action. This work occurred between mid-2009 and mid-2012.

Part of this work was to complete a subsurface investigation at the Grainger site so that the underlying geology could be understood. From this study [4], entitled "Conceptual Site Model and Evaluation of Alternatives for Closeout of Wastewater Ash Ponds – Closeout Plan Phase 1," Santee Cooper learned the following:

> The Site is located in the lower Coastal Plain of South Carolina. The lower Coastal Plain strata consist of a series of unconsolidated sediment and partially lithified sedimentary rocks ranging in age from the Late Cretaceous Period to the Holocene Epoch. Locally, the strata at the site consists of two formations including the Socastee and Bear Bluff Formations (DuBar et. al. 1974). The local Socastee Formation represents the near-surface native unconsolidated material and is composed of an interbedded sequence of sand, silt and clay. The Socastee Formation has been segregated into an upper and a lower unit for purposes of defining the local hydro stratigraphy. The upper Socastee Formation is primarily composed of fine-grained material (silt and clay) with a smaller percentage of sand interbedded throughout. The lower Socastee Formation is composed primarily of well sorted sand that becomes coarser with depth to the contact with the underlying Bear Bluff Formation. This sandy lower portion of the Socastee Formation is the dominantly transmissive geologic unit at the site. The Bear Bluff Formation was encountered at depths varying from 15 to 25-feet below ground surface. The Bear Bluff Formation is described as a very stiff gray clay with trace amounts of very fine sand. Historical borings of the Bear Bluff indicate that the Bear Bluff Formation is at least 35-feet thick. The clay composition of the Bear Bluff Formation is considered a confining unit that greatly restricts or completely inhibits the vertical flow of groundwater from the overlying unit.

Tennessee Valley Authority Kingston Spill

On December 22, 2008, at approximately 1:00 a.m., the northwest side of a dike used to contain coal ash failed at the dewatering area of the Tennessee Valley Authority (TVA) Kingston Fossil Plant, located at 714 Swan Pond Road in Harriman, Roane County, Tennessee. After the dike failed, approximately 5.4 million cubic yards (cy) of coal ash was released into Swan Pond Embayment and three adjacent sloughs, eventually spilling into the main Emory River channel. The release extended approximately 300 acres outside of the fly ash dewatering and storage areas of the plant. EPA's Coal Combustion Residuals Rule (CCR Rule) was promulgated based on this event.

Fig. 10.4 Kingston spill

A graphical representation of the site is shown in Fig. 10.5 [4].

This underlying geology, particularly the high hydraulic conductivities in the Socastee Formation, as well as the fact that fly ash was stored within the Socastee formation, created conditions whereby arsenic and other contaminants could leach from the ash impoundments and enter the groundwater closest to the ground surface (typically called the superficial aquifer) and thus impact the groundwater monitoring wells.

10.5 Pond Closure Options

Decreased power demand related to the economic downturn of 2008–2012 had relegated Grainger Station to a low capacity use facility. From 2008 through 2012, Grainger Station only operated during peak times, typically peak summer heat and the coldest days of winter in order to maintain voltages for reliability purposes in the Grand Strand region. In 2012, Santee Cooper completed a transmission line upgrade that maintained reliability during peak energy use in the Grand Strand area without the need for the load from Grainger Station. For those reasons, Santee Cooper made the decision to idle Grainger in early 2012. In this case, to "idle" a coal plant means to take the plant offline, but it could be brought back online within a few days if power demand dictated it.

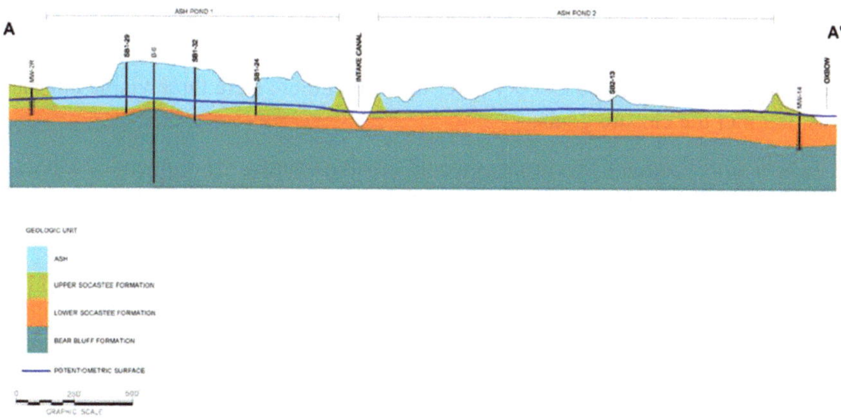

Fig. 10.5 Subsurface representation at Grainger Station

One important modification Santee Cooper made at Grainger Station prior to 2010 was to expand the fly ash silo to accommodate sufficient ash storage for recycling instead of wet sluicing to the ash impoundments. As part of an overall ash management strategy, Santee Cooper had partnered with Southeastern Fly Ash (SEFA) to construct a carbon burn-out facility at Winyah Station in Georgetown, SC. This process would accept high-carbon fly ash from throughout the Santee Cooper system and, utilizing self-sustaining fluidized bed combustion, beneficiate the fly ash to a form suitable for use in ready-mix concrete. Prior to this project, most of the Grainger ash was sluiced to the impoundments as it was unusable for ready-mix concrete. After the silo was expanded, nearly all the fly ash generated at Grainger Station was sent to beneficiation processing then into the ready-mixed concrete market, and little new ash was placed into the impoundments.

Since the station was idled and because there was no need for fly ash impoundments even if the units were to be reactivated because of the increased silo storage to beneficiation, the decision could be made to close at least one and perhaps both ash impoundments. The direct benefit of closure would be mitigation of groundwater issues likely related to the impoundments. Santee Cooper set goals that the closure:

- Must be protective of human health and the environment.
- Must be in accordance with regulatory requirements.
- Must use technically feasible and proven engineering and construction methods.
- Should be cost-effective.

With these goals in place, Santee Cooper's consultants began to develop several alternatives to close the ash impoundments. While numerous options were developed and discussed, Santee Cooper determined that there were three viable alternatives that met the closure goals:

1. *In Situ Stabilization* – this process would involve fixating the impounded ash and water with bentonite or a similar material to the point it became hardened and impermeable. This would minimize leaching and migration of contaminants and leave a foundation suitable for grading and future development.
2. *Capping in Place with Lateral Impermeable Isolation* – this process would surround the impoundment with an impermeable slurry wall and cap the ash with an impermeable surface layer. The groundwater could not move laterally though the impounded ash and rainfall could not penetrate the surface cap. Santee Cooper termed this solution "cap and vault."
3. *Complete Removal of Ash Pond Solids and Off-Site Disposal* – this solution involved removal of the ash over a defined period to a modern, lined industrial or Resource Conservation and Recovery Act (RCRA) Subtitle D landfill. Once removed, the impoundments would be returned to natural wetland habitat.

During the time Santee Cooper was working through the closure alternatives process, two influencing events occurred. First, in mid-2012, a local conservation group filed a lawsuit against Santee Cooper alleging violation of state environmental regulations based on groundwater within the on-site superficial aquifer not meeting drinking water standards. Second, at the end of 2012, Santee Cooper announced the permanent closure of Grainger Station for various reasons, among them reduced loads and upcoming environmental regulations including the Mercury and Air Toxics (MATS) rule that made compliance at Grainger Station impracticable.

In early 2013, Santee Cooper chose option 2, cap and vault, as the impoundment closure methodology. This approach met the impoundment closure goals and was also the lowest-cost option, roughly half the cost of option 3, ash removal and off-site disposal. One primary cost driver of the off-site disposal option was both the capital and operational costs associated with taking up landfill space at any landfill owned by Santee Cooper or others. Santee Cooper filed a cap and vault closure plan with SCDHEC in May 2013.

As part of the closure application process, Santee Cooper conducted public meetings to present the plans to the community and met with local city and county council members and officials regarding the plan. Opinions of the proposal were mixed. While a portion of the community was accepting of the cap and vault proposal, the conservation community clearly desired to have the ash removed from the site for a number of reasons, including removing the ash because it posed an environmental risk and as part of a national campaign to close ash impoundments through ash removal. To give a clearer picture of the opposition to the cap and vault option, at an April 2013 public meeting, Michael Corley with the South Carolina Environmental Law Project and part of the conservation community stated his opposition to the cap and vault option, saying "I believe that horse is so objectionable it needs some additional beating" [5]. In addition, Conway City Councilman

Tom Anderson stated, "We don't need to leave this stuff in downtown Conway at all. Enough said" [5].

As mentioned earlier, Grainger Station is located within the city limits of Conway, SC, and in May 2013, the city council passed a resolution that clarified their preference for an ash removal solution. It was very clear in mid-2013 that Santee Cooper's proposed solution, while technically sound, was going to cause concern to key segments of the public.

10.6 Decision Point: Is There a Better Way?

Santee Cooper had maintained a successful coal combustion byproduct program for over 25 years. Dry fly ash from other stations had been utilized in cement manufacture (three cement kilns are located near Santee Cooper's coal-fired generating stations). As already outlined, this same ash that meets certain specifications had also been marketed directly to concrete ready-mix batch plants. While the national average ash recycling rate was in the 40–50% range [6], Santee Cooper's was far higher, with some years exceeding 80%. This fly ash was segregated from other ash, stored dry in silos, and loaded into sealed pneumatic trucks for transport to the cement and concrete industries. Wet ash stored in impoundments is a different material altogether. This material, typically called "pond ash" contains a combination of both fly and bottom ash. It's moist, and at Grainger, the moisture was high in some areas, approaching 25%. Also, over time, trees, grasses, and shrubs will grow in ponded ash, leaving behind residual biomass from roots. In summary, ponded ash is a hodgepodge of coal combustion residuals, not the consistent product of dry fly ash. Ponded ash is a difficult product to market.

For several years, Santee Cooper had screened pond ash from its ash impoundment at Jefferies Station with the intent of marketing it to cement manufacturers. Screening ponded ash removes bottom ash along with biomass (roots and other plant material). The screened product was stacked in windrows to dewater, and the final product was similar to dry fly ash but with a higher moisture content. Santee Cooper had limited success marketing this material, as cement companies were required to pay a small amount for the material and transport it to their facilities.

As of the summer of 2013, Santee Cooper was still moving forward with the cap and vault solution to the ash impoundments at Grainger Station. As Santee Cooper continued to evaluate stakeholder concerns to remove the ash from the facility, another idea emerged:

> What if we could remove the ash from the impoundments, not to a landfill for disposal, but remove it for recycling for cement manufacture and do it so it had a similar impact to our customers as the low-cost option of cap and vault?

If this could be accomplished, there would be the added benefit of lower long-term risk since the ponded fly ash would be incorporated/encapsulated in the cement product as opposed to undergoing long-term storage within the "cap and vault."

10.7 Changing the Solution: Ash Removal with Recycling

One of the primary cost burdens of removing and landfilling the impounded ash is the tipping fees associated with disposal. Tipping fees would include the cost to operate the landfill and also the capitalized cost of constructing the landfill. Usually these are between $15 and 40/ton of waste material depending on land cost and complexity of construction. By recycling the ash, the tipping fees would not be incurred, making the option more viable.

Santee Cooper's Environmental Management group began to work to develop this solution. The prior experience at Jefferies Station of screening and dewatering the ash for sale to cement manufacturers proved very valuable in that the concept was feasible, but cost and contracts with cement kilns needed to be vetted. In prior years, Santee Cooper would screen and stack fly ash into windrows and allow cement companies to utilize this material for a cost per ton (usually less than $0.25/ton). The cement companies would also be required to transport and load the material onto their trucks. Costs could be significant depending on the distance from the cement kiln to the ash impoundment. Dry fly ash, on the other hand, was much easier to handle. Sealed pneumatic truck-loaded dry fly ash from a silo at a coal plant can be quickly and easily loaded and similarly unloaded into a silo at a cement kiln. The ponded ash process described above was much more involved with sorting, stacking, and transporting in open trucks, with a somewhat less consistent product (Fig. 10.6).

The first hurdle was to work out the issue of understanding the cement kiln preference for dry fly ash over the processed ponded ash. While the ponded material had been deemed technically feasible, it was determined there was a need to provide incentives for the cement kiln to accept the alternative due to pond ash storage issues (stacked outside as the processed pond ash is too wet to be stored in silos) and the additional fuel use in the cement manufacturing process that would be needed to remove the moisture contained in the ponded ash. After much discussion with the kilns, Santee Cooper agreed to excavate, screen, and transport ponded ash to the cement kilns. This involved a distance of approximately 125 miles. The tipping fee savings facilitated Santee Cooper's ability to provide this offer to the cement kilns. Santee Cooper initiated contracts with two cement kilns to provide ponded ash for cement manufacture.

Of course, digging out the impoundments, sorting the ash, loading it, and transporting it to the cement kilns over 100 miles away were no small tasks. Based on Santee Cooper's experience at Jefferies Station, these activities could be handled either internally or with contractors or both. In addition, firm costs for these activities were already known and could be utilized in the final cost analysis. Initially, Santee Cooper anticipated that transport from Grainger Station to the cement kilns in Holly Hill, SC, would be via truck. Estimates were to transport 1000–1500 tons of sorted fly ash per day, which translates to 40–60 truckloads each day. The initial concerns with the excavating, sorting, and transport operation were:

Fig. 10.6 Pond ash sorting and loading

- Dust from digging and screening.
- Traffic issues from the sheer number of trucks entering and exiting the facility and the requirement to travel through downtown Conway, SC, on a daily basis.
- Any air emissions associated with ash removal, in particular dust.

While dust can be controlled through best practices, Santee Cooper made the decision to look into other transport options in anticipation of traffic concerns. The alternatives were:

- *Rail* – A private line rail already ran to the site, which last received coal 2 years prior in 2011. Even during Grainger Station's last coal shipments, it was well-known the rail was in a state of disrepair. Estimates to repair the rail for years of ash removal were in the tens of millions of dollars, ruling out rail as a viable option.
- *Barge* – One thought was to bring a barge up the Waccamaw River, load it with fly ash at the site, and then take it downstream to a staging area to unload from the barge onto trucks for the final trip to the cement kiln. This option was quickly discarded as the cost of multiple handling of the ash and the risk of overwater transport made truck transport much more viable.

After considering these options, truck transport was the only viable method to move ash from the impoundment to the cement kiln. The strategy would then be to work out a schedule suitable to both the city and the surrounding community.

An internal deadline of November 2013 was set to finalize a disposal decision for the Grainger Station ash impoundments. Ultimate disposal options as well as agreements in principle with the cement kilns were developed over a matter of months in order to meet this deadline.

While these agreements were being negotiated (and there was certainly no guarantee an agreement would be reached to recycle the ponded ash), Santee Cooper began to inform stakeholders that solutions beyond the cap and vault proposal were being considered. Stakeholders were engaged to attempt to garner support for the cap and vault solution while also letting them know other options were being considered as long as those options met Santee Cooper's goals for the project, which were that the impoundment closure solution must be technically sound, would protect the environment and was beneficial to customers and the community. Santee Cooper's community relations group stayed engaged with City of Conway and county officials. Others within Santee Cooper, including Environmental Management as well as the legal department maintained a dialogue with the public as well as the conservation community during this time.

By late November 2013, all agreements were in place to remove the ash from the impoundments for transport to the cement kilns for use in cement manufacture. At that point, the external stakeholders were informed of the decision to remove the ash from the impoundments. The decision was well received within the community. This desirable solution was accomplished by internal collaboration as well as by engaging external stakeholders.

Since the ash was to be removed from the impoundments, this provided Santee Cooper a path forward to settle all legal matters brought by conservation groups over groundwater contamination. In order to dismiss these suits, Santee Cooper agreed to:

- Withdraw the cap and vault closure plan previously submitted, and resubmit a plan for ash removal from the site.
- Complete ash removal at Grainger Station no later than December 2023 and make a good faith effort to finish by December 2020.
- Complete closure with a plan approved by SCDHEC.

In response to the settlement, Frank Holloman, an attorney with the Southern Environmental Law Center stated, "This is an historic agreement that removes toxic coal ash from beside the Waccamaw River and from Conway. This settlement is good for Conway, for the River, and for Santee Cooper, and we thank Santee Cooper for reaching this agreement" [7]. Christine Ellis, the Waccamaw River keeper stated, "This settlement provides for the protection of our beautiful black water Waccamaw River. On behalf of our members and supporters, and our community as a whole, we are grateful to Santee Cooper for agreeing to remove its toxic coal ash and helping us to achieve our goal of fishable, swimmable and drinkable water for our families and our future. This is a great day for the Waccamaw River and for Conway, our Rivertown" [7].

2013 estimates to complete the cap and vault closure plan were in the range of \$27–43/ton of ash. 2013 ash removal with recycling estimates were approximately \$43/ton of ash but with the added benefit of risk reduction as the ash was to be completely removed from the site.

Since the closure plan was approved by SCDHEC in mid-2014, Santee Cooper has been removing ash and transporting 30–40 loads each weekday by truck to the contracted cement kilns in Holly Hill, South Carolina.

10.8 Conclusion

When Santee Cooper began this process, the company was well aware there were different ways to deal with ash impoundment closure. Santee Cooper's first focus, as it should be, is to consider what is best for the customer and which is the lowest-cost option, and in this case, the best option was the "cap and vault" solution. Working through the closure process, Santee Cooper learned of other stakeholder and customer viewpoints about what was considered the "best" solution. Each of these solutions had an associated trade-off, whether it was cost, environmental impacts or future risk. Santee Cooper was able to consider all these viewpoints and weigh the trade-offs. Ultimately, Santee Cooper implemented a closure methodology that minimized risk and that satisfied environmental protections of the surrounding ecosystem, regulatory agencies, and the local community with little downside to what was originally believed to be the primary driver, the lowest customer cost.

As of December 2017, 3.5 years into the project, Santee Cooper has removed over one million tons of ash from the existing impoundments at Grainger Station, with the majority of this ash recycled into cement manufacture. Small quantities of sorted organics were disposed of in appropriate landfills.

Arsenic concentrations in groundwater have changed significantly in the monitoring wells at the site. The wells with the highest arsenic concentration were those between the impoundments and the Waccamaw River, with historical highs at over 3000 ppb arsenic. Recent sampling events have shown arsenic reductions of between 60% and 95%. It is anticipated this trend will continue as the ash continues to be removed.

As of the end of 2016, the current all-in cost of removing, processing, and transporting the ash to cement manufacturers is approximately $27/ton of ash removed. This equates to the lowest-cost estimate for the cap and vault solution proposed in 2013.

In summary, the ash is being removed from the impoundments for approximately the same cost as the low-cost alternative but with the added benefit of risk reduction because the ash is being used in cement, and groundwater quality is improving as a result. When asked in 2015 by the Charlotte Business Journal what was his most significant accomplishment at Santee Cooper, then-CEO Lonnie Carter replied, "If I picked one thing, I would say getting our beneficial reuse program to recycle pond ash up and running" [8].

Author Profile

Jay Hudson (Retired) Hudson was the administrator of Environmental Management at Santee Cooper. Mr. Hudson's over-30-year professional career has been focused on various environmental engineering issues including air, water, and waste. Through his position at Santee Cooper, Hudson served on the American Public Power Association's Clean Air Task Force and New Generation Task Force as well as the Utility Water Group Policy Committee. From 2008 through 2014, Hudson served as Chairman of the Large Public Power Council's Environmental Task Force guiding environmental policy for over 20% of US electricity customers. Hudson has published several articles on pollution topics, is a member of the American Chemical Society, and is a registered Professional Engineer in South Carolina. Mr. Hudson holds a Bachelor's Degree in Chemical Engineering and a Master of Engineering degree from the University of South Carolina with an emphasis in Water Resources and Environmental Engineering.

Company Profile

Santee Cooper is South Carolina's state-owned electric and water utility and the state's largest power producer, supplying electricity to more than 176,000 retail customers in Berkeley, Georgetown, and Horry counties; as well as to 26 large industrial facilities; Central Electric Power Cooperative; and two municipal electric systems, the cities of Bamberg and Georgetown. Santee Cooper also generates the power distributed by the state's 20 electric cooperatives to more than 771,000 customers in all 46 counties. Approximately 2 million South Carolinians receive their power directly or indirectly from Santee Cooper. The utility also provides wholesale water to consumers in Berkeley and Dorchester counties and the towns of Santee and Bowman.

References

1. "Coal Ash (Coal Combustion Residuals, or CCR)," United States Environmental Protection Agency, https://www.epa.gov/coalash. Accessed 15 December 2017
2. "Frequent Questions about the Coal Ash Disposal Rule," United States Environmental Protection Agency, https://www.epa.gov/coalash/frequent-questions-about-coal-ash-disposal-rule#4. Accessed 15 December 2017
3. "Steam Electric Key Waste Streams," United States Environmental Protection Agency, https://www.epa.gov/sites/production/files/2015-05/steam-electric-key-wastestreams_2025x1571.png. Accessed 15 December 2017
4. ARCADIS. "Conceptual Site Model and Evaluation of Alternatives for Closeout of Wastewater Ash Ponds – Closeout Plan Phase 1." 17 March 2013. Public document prepared for Santee Cooper and submitted to South Carolina Department of Health and Environmental Control, 18 March 2013
5. "Public calls for coal ash to be hauled away from Conway's Grainger plant," *Sun News*, 23 April 2013, http://www.myrtlebeachonline.com/news/local/article16646558.html. Accessed 15 December 2017
6. "2010 Coal Combustion Product (CCP) Production & Use Survey Report," American Coal Ash Association, October 2011, https://www.acaa-usa.org/Portals/9/Files/PDFs/2010_CCP_Survey_FINAL_102011.pdf. Accessed 15 December 2017
7. "Santee Cooper Agrees to Remove Coal Ash from the Waccamaw River and Conway," Southern Environmental Law Center press release, 19 November 2013, https://www.southernenvironment.org/news-and-press/press-releases/santee-cooper-agrees-to-remove-coal-ash-from-the-waccamaw-river-and-conway. Accessed 15 December 2017
8. "Santee Cooper Making Headway on Carbon Emissions," *Charlotte Business Journal,* 22 May 2015, https://www.bizjournals.com/charlotte/print-edition/2015/05/22/santee-cooper-making-headway-on-carbon-emissions.html?s=print. Accessed 20 January 2018

Chapter 11
Creating Success for TVA Through Sustainable Solutions and Customer Collaborations

Monte Lee Matthews, Dawn Best, Laura Doze, Todd Kiefer, Madhu Ranade, Karen Utt, and Rob Wise

Abstract The Tennessee Valley Authority (TVA) is the largest publicly owned power company in the United States, having originated from a Congressional mandate in 1933 as a solution to the flooding, poverty, and lack of electricity in much of the Tennessee Valley. Learning to strike the appropriate balance between energy production needs, flood control, economic development, and stewardship often involves competing forces that require TVA and its customers to be innovative and collaborative. Maintaining a balance now and in the future is sustainability at TVA. However, sustainability does not necessarily involve a tradeoff in the traditional sense where companies must make a choice between seemingly incompatible ideas. Instead, TVA's business decisions, strong customer relationships, and customer-oriented programs represent a mutually beneficial business agreement. Sustainability is smart business. This chapter delves into sustainability at TVA and the dynamics of three-customer relationships to highlight how collaboration leads to sustainable solutions. In fact, TVA's efforts have helped recruit or expand the operations of 218 companies in fiscal year 2017. This resulted in attracting and retaining more than 70,000 jobs and $8.3 billion in capital investment to the region.

M. L. Matthews (✉) · D. Best · L. Doze · K. Utt
Tennessee Valley Authority, Knoxville, TN, USA
e-mail: mlmatthews@tva.gov

T. Kiefer
East Mississippi Electric Power Association, Meridian, MS, USA

M. Ranade
Steel Dynamics, Inc., Fort Wayne, IN, USA

R. Wise
Resolute Forest Products, Montreal, QC, USA

J. Fox, M. Scott (eds.), *Sustainable Electricity II*,
https://doi.org/10.1007/978-3-319-95696-1_11

11.1 Introduction

When President Franklin D. Roosevelt signed the Tennessee Valley Authority (TVA) Act on May 18, 1933 [1], it marked the end of a 14-year effort. The ceremony in Washington that day did not take long, but the struggle that preceded it had been anything but brief.

There was constitutional justification for the TVA Act, founded upon the federal government's right to improve navigation and flood control. But the roots of the TVA "idea" had grown from the most prominent issues of public policy during those years between the World Wars – the production and distribution of hydroelectric power, resource conservation and development, and regional planning. In addition to these responsibilities, the new agency was given a broad mandate to promote the economic and social well-being of the people of the region [2]. Today, we call this sustainability.

TVA considers sustainability in overall planning and business decisions. In fact, TVA goals often align with our customer goals – to provide low-cost and clean power that helps Valley businesses prosper. Much planning goes into the business of sustainability in order for TVA and its customers to thrive. Sustainability does not happen by accident. In the 2015 Integrated Resource Plan (IRP) [3], TVA conducted an in-depth analysis of the Tennessee Valley's long-term energy needs. "The IRP discusses ways that TVA can meet future electricity demand economically while supporting TVA's equally important mandates for environmental stewardship and economic development across the Valley" [3]. The various resource options were weighed against criteria that reflect TVA values and a sustainable future. These criteria include cost, risk, environmental stewardship, system flexibility, and Valley economics. The results of the effort concluded that a diverse generation portfolio (nuclear, coal, gas, hydroelectric, and renewables) is the best way to optimize both TVA values and the values of our customers and the community. This demonstrates the importance of understanding the core values of stakeholders to manage traditional tradeoff decisions.

The idea of sustainability as reflected in the Integrated Resource Plan supports TVA's mission—energy, environmental stewardship, and economic development (the 3 E's). The 3 E's also echo our customers' business needs and are critical to the ongoing economic strength in the Tennessee Valley (Fig. 11.1).

Brenda Brickhouse, TVA's Vice President of Environment and Energy Policy and Chief Sustainability Officer, describes how the 3 E's interact: "TVA provides cleaner, reliable and low-cost energy to support sustainable economic growth in the Tennessee Valley and to engage in proactive environmental stewardship in a balanced and ecologically sound manner" [4].

Sustainability is *smart business* for TVA and its customers. And, sustainability does not necessarily involve a tradeoff in the traditional sense where companies must make a choice between seemingly incompatible ideas. Instead, TVA's business decisions and customer programs, such as time-of-use (TOU) rates and energy efficiency, represent a mutually beneficial business agreement that forms the basis of community and company sustainability. TVA's commitment to meeting customer needs has

Fig. 11.1 TVA's 3 E's

resulted in recruitment and expansion of 218 companies throughout the Valley in 2017, attracting and retaining more than 70,000 jobs and $8.3 billion in capital investment to the region [5]. This is TVA's story of collaboration and sustainability.

11.2 From Then to Now

TVA has made difficult decisions throughout our history that have impacted livelihoods, businesses, and the Valley economy both positively and negatively and have also required money to enable sustainable changes. The money in the early days came from federal funding; money in more recent days comes from TVA's ratepayers.

Continual sustainable change is not without much consideration. How does TVA improve the economic vitality of a region? How much money does TVA invest? Does TVA improve the river system for flood control even though it may impact some Valley residents? Is it appropriate to burden some now while future generations may benefit? These are some of the questions and decisions that TVA has wrestled with over the years.

In 1933, personal income in the TVA Valley's region averaged $168 per capita, 45% of the national level. By 2011, because tough sustainable decisions were made, personal income was $34,442 per capita, 83% of the national level – approximately 201 times what it was in 1933 – while the national average multiplied about 110 times over the same period [6]. Likewise, the Tennessee Valley has avoided billions of dollars in economic impact through TVA's decision to invest in flood control. Approximately $17 billion in flood damage has been averted in Chattanooga, Tennessee alone since 1936 [6].

The choices that TVA and our customers and stakeholders make today are ultimately important for continued economic growth, for enabling a better place to live, and for the continued generation of sustainable energy in the Valley.

11.2.1 Economic Development and Sustainable Growth in the Tennessee Valley

TVA's commitment to sustainability is important to economic development in the Valley as more businesses demonstrate a commitment to increasing sustainability efforts (Fig. 11.2).

Communities and regions that embrace sustainability, especially carbon dioxide (CO_2) emissions reduction, are beginning to be viewed as more progressive. It differentiates those communities and gives them a competitive advantage with new businesses seeking an area to locate.

11.3 Sustainable Planning for Customers

TVA is proud of the work it does related to carbon competitiveness for electricity. Some utilities may have lower CO_2 emission rates but higher prices, and utilities with lower prices may have higher CO_2 emission rates. When looking at these factors, TVA is well-positioned for both. Currently, TVA has the second lowest industrial rates among regional utilities, and TVA's carbon content is also in a top competitive position. Companies looking to relocate to the Valley or expand current business often find this attractive. For example, TVA's calendar year 2017

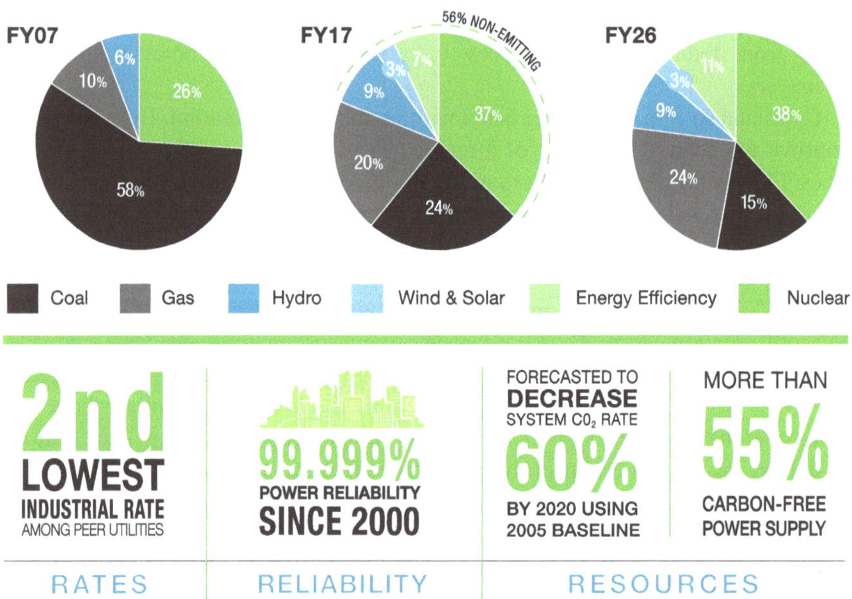

Fig. 11.2 TVA's carbon and economic development impact

as-delivered system CO_2 emission rate was 850.91 pounds per megawatt hour (lbs/MWh), including emissions and generation from owned and purchased power. The most up-to-date Environmental Protection Agency's national and regional carbon rate is 998.40 and 1185.40 lbs./MWh, respectively [7]. Therefore, TVA's carbon rate is more than 25% lower than the regional rate, and the TVA carbon rate is projected to lower by about 60% below 2005 levels by 2020.[1]

While the carbon footprint is important to current and prospective customers, TVA has found that it is also critical to working with and incentivizing communities across TVA's service area to be even more sustainable. The Valley Sustainable Communities Program, developed by TVA's economic development group, provides a framework and assistance to communities that are committed to learning and deploying sustainable practices. Practices include sustainability planning, energy efficiency, water conservation, climate protection, and green design and construction. Communities are ranked (platinum, gold, and silver) on their level of sustainable practice in 15 categories. As these community standards are raised, the communities become even more attractive to prospective business, and their growth potential increases [8].

TVA conducts annual customer surveys, which coupled with insight from prospective customers indicate that reliable and low-cost electricity, along with a strong focus on sustainability, especially carbon footprint, are two of the most important factors attracting customers to the TVA territory. TVA intentionally reduced coal-fired generation on its system while shifting to more nuclear, natural gas, and renewable generation sources. These factors, combined with TVA's stewardship of the area's natural resources and the Tennessee River system, make the Tennessee Valley an attractive area to call home.

TVA offers a variety of programs and services for enhancing sustainable performance. Energy efficiency is an important program that helps companies better manage their energy. TVA also offers companies the option of purchasing Renewable Energy Certificates (RECs), which help to offset a company's CO_2 emissions.

As TVA plans program offerings and implements business decisions, there are always challenges and considerations. How many dollars should TVA invest in energy efficiency technology, clean energy technology, or new renewable generation? How does a new energy source become efficiently integrated into our energy mix and still be able to optimize our transmission system? How does a decision mesh with prospective customer needs and perspectives? Many of these questions raise the issue of tradeoff considerations. Because of the complexity of decisions and the realities of funding, it becomes critical for all of TVA's decisions to be made through our business planning process, which takes a long-term, sustainable, and customer-focused approach. TVA views these programmatic decisions as investments to attract businesses and industries who share an ever-increasing focus on sustainability. Since 1977, TVA has spent more than $6.7 billion in clean air technology, achieving a 98% reduction in sulfur dioxide emissions and a 94%

[1] Marketing & Communications chart representing Environment & Energy Policy statistics, Tennessee Valley Authority, November 2016.

reduction in nitrogen oxide emissions. TVA is continuing to reduce its carbon footprint even further [9]. Since 2010, energy efficiency initiatives by TVA and local power companies have reduced electricity consumption by 765 gigawatt-hours, which is the equivalent energy to power 50,000 area households for an entire year. These results were achieved through wise planning and provide tangible evidence to future customers that TVA incorporates a sustainable focus [6].

11.3.1 Trends in Customer Carbon Solutions

The challenge to TVA and other utilities is to have the right asset mix at the lowest possible cost when distributed energy resource solutions are deployed. The resources chosen must keep the grid properly in sync, instantaneously balancing supply and demand, and maintain the standard frequency and voltage needed to avoid blackouts.

Utilities are beginning to weigh in on how best to meet this evolving customer interest in renewable energy. Outside of the existing competitive electricity markets, utility "green pricing" programs have typically provided unbundled RECs that are separate from energy. Green tariffs, or riders, are an emerging option to markets where there is no opportunity to access fixed-price renewable energy. These programs allow eligible customers to buy bundled renewable energy and REC products and cater to customers' preferences for a more direct financial connection to nearby renewable energy projects. Purchasing RECs allows a company to further reduce its carbon footprint. Ultimately, this offers customers a product with greater economic value than unbundled RECs alone.

New investment in renewable energy still outweighs new investment in fossil fuels, and corporations, which account for around half of global electricity consumption, are creating new markets for renewable energy. Six hundred forty-four businesses and investors with more than $15.5 trillion in market cap have made commitments through *We Mean Business*, a coalition of organizations working to foster and showcase bold business action on climate change [10]. Many companies focused on sustainable improvement are joining a set of campaigns developed by The Climate Group such as RE100 (renewable power), EP100 (energy productivity), and EV100 (electric vehicles). Many of TVA's customers and/or their parent companies are members of these or similar climate change coalitions and the sustainable improvement campaigns.

Inquiries by customers around sustainability, in particular carbon, have trended upward over the past several years. Businesses and investors are increasingly aware of climate-related risks, whether the impacts are physical, financial, reputational, or regulatory. The amount of corporate disclosure of environmental data continued to grow in 2016. The State of Green Business 2017 reported four in five (81%) S&P 500 companies issued sustainability reports [11]. The year 2016 also saw a rapid increase in stock exchange disclosure requirements, with 38 exchanges around the world now requiring some sort of environmental, social, and governance disclosure from companies as a condition for listing [11]. The most recent CDP report similarly found that two out of three companies disclose GHG emissions and this disclosure trend is likely to continue rising [12]. The CDP report also found that 85% of companies reporting have already set absolute and/or intensity targets to reduce their greenhouse gas (GHG) emissions. While 56% of companies have set targets for 2020 and beyond, only 14% have set targets for 2030 and beyond [12]. This trend is likely to continue, given the added pressure created by the Paris Agreement on climate change. Our customers' recent trends in renewable generation and carbon emissions transparency indicate a growing need for TVA as a power supplier to also meet these evolving needs.

11.3.2 Customer Needs and Carbon Competitiveness

Many of TVA's directly served customers as well as commercial and industrial customers of local power companies have annual and long-term CO_2 goals and reporting requirements. This means that TVA customers increasingly want to know their carbon footprint based on the electricity purchased from TVA. Many companies disclose their carbon footprints each year, and it is not unusual for corporate buyers or investors to consider carbon performance as part of their decision-making process. Concerned about climate change, companies are competing more often on the size of their carbon footprint. In an effort to give Valley companies a competitive edge, TVA has done groundbreaking work in the area of carbon accounting.

Many companies believe a direct link exists between improved carbon performance and the ability to increase earnings and attract investment [13, 14]. This linkage is driving an increasing number of companies choosing to participate in various sustainability indices, such as the Dow Jones Sustainability Index (DJSI) or FTSE4Good to track their ratings.

In order to calculate a carbon footprint from electricity consumption, a customer will request the CO_2 emissions rate from their utility company. Typically, a utility provides an overall system CO_2 lbs/MWh rate to their customers by simply adding up all CO_2 emitted and dividing by all megawatt hours delivered. TVA learned that some customers want a more precise carbon emission rate—one that is specific to their own usage.

TVA uses a utility cost allocation model to calculate customer-specific CO_2 rates. Like financial accounting, there are also generally accepted standards for carbon accounting. While there has been a North American standard on how to calculate electricity customers' CO_2 lbs/MWh rates since 2009, variations in international calculations were impacting customer competitiveness [15]. TVA and other stakeholders participated in the World Resources Institute's multi-year process to create a new global standard on carbon accounting that is consistent with existing North American practices. This global carbon standard now allows Valley companies with international operations to compete on a level playing field with sister companies outside of North America.

These changes to carbon accounting have allowed TVA to better demonstrate our carbon competitiveness, which supports economic development activities through customer attraction and retention to the region. Customer companies that offer a competitive product price point and an attractive carbon number can more easily make a case for expanding a facility or locating a new plant in the Valley.

11.4 Customer Stories

While many large commercial and industrial power customers factor carbon emissions into business planning, there is also the worry that potential carbon regulations and requirements will put upward pressure on electric rates. A growing number of utilities have begun to recognize the competitive advantage associated with a reduced carbon footprint. Along with attractive electric rates, utilities are leveraging emission reductions as a way of competing for customers.

TVA has a great carbon reduction story, as it has reduced CO_2 emissions 47% below 2005 levels and is projected to reduce CO_2 emissions about 60% below 2005 levels by 2020 (Fig. 11.3). TVA's asset decisions, particularly the addition of 1100 MWs of new nuclear, provide customers with confidence in TVA's lower carbon future.

The following stories highlight how specific companies within the Tennessee Valley region have worked with TVA to optimize short- and long-term sustainable solutions for our customers as we continue our commitment to economic development and stability.

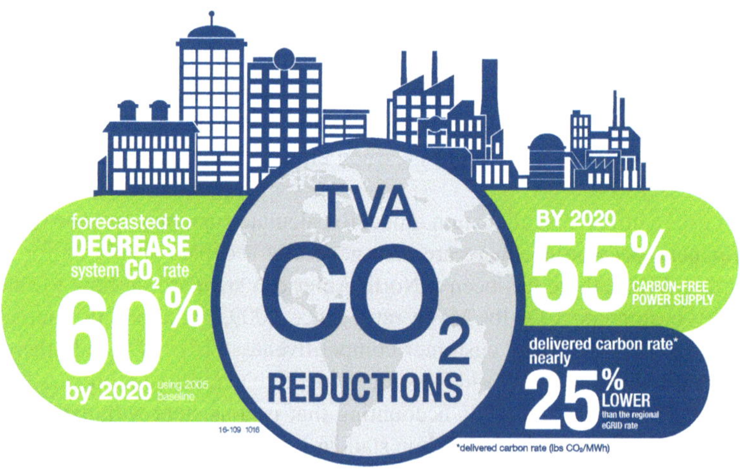

Fig. 11.3 TVA carbon reductions

11.4.1 Steel Dynamics, Inc.

Steel mills consume vast amounts of electricity, often driven by equipment such as electric arc furnaces (EAF) that require an on-demand, uninterrupted power supply to operate effectively. This consumption is often so large that electric utilities take the impact of EAFs into consideration in their planning and day-to-day operations. Steel Dynamics, Inc. (SDI) is one of the largest domestic steel producers and metal recyclers in the United States, with facilities located throughout the United States and Mexico.

TVA and SDI have worked together to explore and gain mutual benefits between TVA ratepayers and SDI's industry-leading Columbus Flat Roll Division (SDI Columbus). SDI Columbus has changed the landscape of Columbus by providing high-paying jobs, focusing on safety, recycling, energy-efficient production, and their dedication to the community (Fig. 11.4).

SDI Columbus is a 3.4 million ton flat roll steel mill employing over 800 people, located in the Golden Triangle Industrial Zone. SDI Columbus is one of 58 energy-intensive large industrial customers that take their power needs directly from TVA. Electricity is the only input in EAF steelmaking that is received and consumed instantly, unlike other production inputs. Hence, a reliable and continuous supply of power is an area of focus for steel producers as well as the electric utilities

Fig. 11.4 Steel Dynamics, Inc. Columbus

that must consider consumer benefit and power grid reliability. And TVA has delivered high reliability since plant start-up.

TVA and SDI Columbus invested sufficiently to ensure that a large load comprised predominantly of electric arc furnaces would not negatively impact other customers' power quality. Together, the investments of SDI Columbus and TVA have provided for a successful integration into the existing power system.

TVA partnered with SDI Columbus on assessments and incentives that made several large energy efficiency projects a reality. Some of these projects included improving energy efficiency with variable speed pumps and drives and improvements to environmental processes. These projects saved enough energy to fully power the plant for 2 weeks or power 3700 houses for an entire year.

Vice President and General Manager Madhu Ranade describes SDI Columbus' relationship with TVA as symbiotic: "I am very impressed with TVA's willingness to listen, provide great economic development programs and partner on energy efficiency initiatives. The value they place on us as a partner helps us reduce our energy consumption and allows power availability for new industrial users without building additional generation capacity. It is customer service at its best" [16].

SDI Columbus has consistently been a leader in energy-saving initiatives and has achieved much success at the Columbus facility. These sustainable solutions were recognized by TVA with a Top Energy Performer Award in 2016.

SDI Columbus has also received a competitive grant from TVA, providing gap money to assist when it's competing with another company outside the Valley for an expansion or capital dollars. These incentives are small in comparison to the total project investment or the value that SDI Columbus provides to TVA, which benefits all customers.

SDI Columbus has participated in TVA's performance-enhanced rates and has helped TVA by reducing load during peak hours and increasing load during off-peak hours. TVA and all customers benefit from this effort because it keeps costs lower during on-peak periods and helps TVA with turndown issues during off-peak periods. Some of the benefits to SDI Columbus include lower power costs, improved energy efficiency, and a lower carbon footprint.

In 2015, SDI Columbus switched to the new and improved time-of-use (TOU) rate that it and other customers helped TVA develop through TVA's strategic pricing initiative. This rate is designed to help both TVA and customers reduce costs.

The TOU rate structure follows a rate schedule that varies according to months, seasons, and designated times throughout the year. The TOU structure provides customers an opportunity to lower overall cost by incentivizing off-peak energy usage. TVA has seen a drop in fuel, maintenance, and operation costs, and SDI Columbus has realized a drop in power costs. SDI Columbus's price per kWh has dropped every year since 2013 due to the partnership between SDI Columbus and TVA. It is clear that working together produces great benefits to all involved.

Like most initiatives worth doing, there were tradeoffs and impacts to SDI Columbus in the pursuit of mutual benefit. SDI Columbus collaborated on and participated in some of TVA's interruptible power (IP) products. These products provide TVA a cost-effective way to maintain 99.999% reliability. Most plants like SDI

Columbus are designed to run at full capacity, so the company modified its production practices to preserve quality and safety. This required investments in training and operations. The energy efficiency improvements also meant diverting capital dollars away from other options.

The dedicated team at SDI Columbus made these initiatives work, and the results continue to be favorable. Working together with TVA provided a sound approach to implementing workable energy efficiency projects and lowering power costs.

SDI's continued focus on energy-saving projects is not only beneficial to the company but also benefits the immediate community through job creation and growth. The reduction in power consumption returns capacity to future customers without any capital investment to the power producer and supplier. It also reduces impact on the environment by reducing carbon emissions and conserving natural resources for future generations.

SDI Columbus and TVA continue to partner and strengthen their business relationship by collaborating on economic development, electric product options and offerings, rates, transmission system improvements, and technical energy assessments and projects. Our work together is just one step forward for others to follow, and there are many more opportunities to reduce energy consumption, save money, protect the environment, and conserve natural resources.

11.4.2 Naval Air Station Meridian, East Mississippi Electric Power Association, and TVA

East Mississippi Electric Power Association (EMEPA) was only 20 years old when the first shovel of earth was thrown to begin construction of what is now known as Naval Air Station Meridian. For nearly 59 years, this base has been served by EMEPA delivering electricity generated by TVA (Fig. 11.5). This relationship has been a dynamic one.

Naval Air Station Meridian straddles Lauderdale and Kemper Counties and is about 15 miles north of Meridian, MS. Meridian has a population of 40,000 (base population comprises 2000, with about 600 active duty military living on the base). With NAS Meridian, Meridian is the sixth largest city by population in Mississippi. NAS Meridian's economic impact to the region is greater than $200 million annually.

Historically, the US military maintained its own electrical infrastructure, and EMEPA's role toward NAS Meridian was straight power delivery. However, that changed in the 1990s when the military began to outsource noncombat-related functions to contractors. In this decade, deregulation swept the power industry, and a profusion of rate options emerged. NAS Meridian began to look to EMEPA for maintenance and consulting services. One of the first projects was replacing two aging switching stations with modernized substations.

Fig. 11.5 Naval Air Station Meridian

The Department of the Navy has aggressive goals aimed at transforming the Navy's energy use. In 1998, EMEPA leadership devised a way to translate TVA's new rates into a benefit for both the Navy and TVA. The base was transitioned from firm power to an interruptible power rate that offered significant cost savings. These savings were used to pay for the installation, operations, and maintenance of 9 MW of TVA diesel generation, sufficient to power the entire base, including an energy management system and controls that permitted operation in island mode. Having an interruptible power rate does mean that there is a chance that TVA may have to interrupt the power. TVA indeed has had system operational situations arise that have required the base to be interrupted. The local distributer of power, EMEPA, has also taken scheduled line outages to complete line maintenance work. When the line is interrupted or taken out of service for maintenance, TVA utilizes pre-planned switching orders and coordinates with EMEPA to bring the generators online to pick up the base in island mode. EMEPA verifies the local base breakers are opened before bringing the generators online, and then the breakers are closed systematically to pick up the load.

At no additional cost, NAS Meridian has had its own microgrid for 18 years, and this backup has often proved its worth by keeping the base up and running during grid outages caused by trees, animals, tornadoes, and Hurricane Katrina. TVA also benefited by being able to dispatch the generators in grid-tied mode as a peak-shaving plant. TVA saved more than the cost of the generators the very first year of peaking operations. This project was a win for all parties, and NAS Meridian remains today one of only two bases in the Department of the Navy with full on-site backup [17].

As the Navy became more interested in energy efficiency, TVA supported the base by offering engineering consultants to develop and bid plans for upgrades to heating, ventilation, and air conditioning (HVAC) controls and lighting. EMEPA contracted the projects and was reimbursed by the Navy. All told, NAS Meridian has spent more than ten million dollars on energy-efficient improvement projects (directly and via EMEPA), with each one delivering a simple payback of 10 years or less. NAS peak demand has been reduced from 9.5 MW to 5.7 MW. While this has reduced EMEPA and TVA power sales, it has increased the value of the base to the Navy and the nation and helped to insure its continued viability during a climate of budget cuts and base closures. It is the nature of member-owned cooperatives to operate in the interests of their members and to judge returns over multi-decade timeframes. The Naval Air Station is a member whose considerable patronage helped underwrite the growth of the system. The legacy of trust built up between the parties over the decades because of actions like these was likely instrumental in the Navy's decision to privatize the electrical plant to EMEPA in 2008, with proven mutual benefit. In a climate of federal fiscal austerity and basic rest-activity cycles (BRAC), anything that makes NAS Meridian more cost-competitive and energy resilient than other Navy bases with similar missions is in the long-term interest of the community and members that EMEPA serves.

In another pioneering move, so far accomplished at only a handful of bases, the Navy privatized the electrical infrastructure of NAS Meridian to EMEPA in 2009. The Navy at large benefits by no longer having to maintain electrical plant expertise in its ranks. The base benefits from having a dedicated facilities coordinator who can call upon the resources of a full power company.

The latest chapter in this sustainable partnership is solar power. To satisfy twin goals of promoting renewable energy and increasing resilience of power at its bases, the Navy Renewable Energy Program Office approached EMEPA and TVA with a proposal to lease land on the base for construction of a utility-scale solar array. TVA also has a goal of promoting renewable energy, and EMEPA successfully competed for two TVA programs that offered local power companies generous wholesale power purchase rates to offtake solar power. Before the end of 2018, construction will begin on a 6 MWdc PV solar array that will be integrated with the existing base microgrid, so it can generate when the base is islanded due to a grid outage and supplement the diesel generator power to extend their unrefueled run-time.

TVA, EMEPA, and NAS Meridian have a unique and enduring partnership that has evolved over time to capitalize on changing rules, rates, and roles to the mutual benefit of the three parties, the communities we serve, and the nation as a whole.

11.4.3 Resolute Forest Products

Resolute Forest Products is a global leader in the forest products industry with a diverse range of products, including market pulp, tissue, wood products, newsprint, and specialty papers, which are marketed in over 70 countries. The company owns or operates some 40 manufacturing facilities, as well as power generation assets in the United States and Canada. Resolute's corporate headquarters is located in Montreal, Canada, and the company owns a newsprint mill in Grenada, Mississippi that TVA serves directly (Fig. 11.6). Resolute has received regional, North American, and global recognition for its leadership in corporate social responsibility and sustainable development, including a third party certifying 100% of its managed woodlands to internationally recognized sustainable forest management standards. Resolute's sustainability strategy has been developed to align the company's efforts in environmental stewardship and social responsibility with its business strategy. This approach underpins Resolute's vision that profitability and sustainability drive its future. The company seeks to balance environmental, social, and economic performance to enhance its competitive position. Resolute's sustainability strategy has three primary areas of focus: (1) reinforcing its environmental credentials, (2) positioning the company as a competitive employer by attracting and retaining the best employees, and (3) building solid relations in its operating communities. Resolute's

Fig. 11.6 Resolute Forest Products Grenada Mill

Grenada newsprint mill competes in a global market and goes head-to-head with other company operations for capital dollars and expansion opportunities. The primary focus of TVA's partnership with the Grenada mill has been to reduce costs by using available rate structures and programs and through investments in energy efficiency and process improvements. Safety, reduction of carbon footprint, and other sustainability objectives are also critical.

One success story is an energy efficiency improvement project that allowed Resolute to move some load to the off-peak hours and improved the efficiency of their production without sacrificing product quality, resulting in annual energy savings that are the equivalent to supplying power to 5300 homes for 1 year.

With the newsprint business in a declining market due to the expansion of digital technology, the demand for environmentally responsible products is increasing. These factors, combined with Resolute's commitment to sustainability, have motivated the company to offer low-cost, environmentally responsible newsprint.

For the Grenada mill, reducing power costs while maintaining product quality and meeting Resolute's sustainability goals has been a great challenge. The ability to reduce consumption during TVA's peak times affords Resolute lower overall rates as well as participation in TVA's Economic Development Programs and Energy Right® Programs. The benefits the Grenada mill has realized from projects with TVA include lower energy consumption, greater operating flexibility, lower costs, and increased capital investments. They also contributed to a 25% reduction in the mill's absolute scope 2 GHG emissions over a 2012 baseline. As a result of those efforts, TVA recognized Resolute's Grenada mill with a Top Energy Performer Award in 2016 and a Leadership Circle Award in 2017. The Leadership Circle Award is presented to the top three TVA directly served customers with the lowest carbon emission rates.

Tradeoffs include some increases in maintenance costs, changes in production processes, and the need for additional employee training to manage the ramping up and down of pulp production during peak and off-peak hours. Rob Wise, general manager of the Grenada mill, said: "The programs TVA offers, its quality customer service and our partnership are an important part of why we are competitive today." "Both Resolute and TVA strive for sustainability, employee safety, product quality, excellent customer service and low power costs," added Wise. "We also enjoy strong support from local business and community leaders and state government" [18].

11.5 Conclusion: TVA and Customers Working Toward a Sustainable Future

Through careful operational decisions and programmatic efforts, TVA continues to strive for sustainable decisions while managing the river system, providing reliable energy, and growing the Valley's economy – all to improve the quality of life in the Tennessee Valley. Learning to strike the appropriate balance between TVA's energy production needs, stewardship mission, and customer requirements with often competing demands requires innovation, collaboration, and commitment. Maintaining the balance may not be easy, but it is the right thing to do for the people of the region. Sustainable performance does not necessarily involve a tradeoff in the traditional sense where companies must make a choice between seemingly incompatible ideas. Instead, TVA's business decisions, strong customer relationships, and customer-oriented programs represent a mutually beneficial business agreement. TVA recognizes that many customers value sustainable performance as a business requirement. Working with TVA has been an integral part of their success. TVA's customers provide a large percentage of the jobs in the Valley and are among the highest-paying companies in the region. These customers also invest millions of dollars each year in local facilities, bringing additional jobs to the area.

The customer examples presented demonstrate several common issues that required tradeoff discussions to achieve sustainable performance. TVA and each customer had significant discussions focused on how to wisely partner and how to be most effective working together. The following are the common items and the tradeoff discussions which led to mutually agreeable decisions:

- Advance corporate carbon accounting and sustainable decisions. Often customers used regional or national carbon numbers, which were worse than the Tennessee Valley's carbon numbers. Our customers found that TVA could help them produce more exact carbon data which in turn helped them establish better business cases for expansion. TVA, in turn, became more aware of competitive and international carbon standards and expectations.
- Keep monetary rates as low as feasible while also meeting company sustainability goals. In our customer partnerships, there are negotiation sessions to provide the best rate plausible, taking into consideration all aspects of each customer.
- Improve facilities to be more energy efficient, and discuss the best approach for redesigning or maintaining a facility. Sometimes modifying an existing facility is the best long-term solution.
- Manage future risk. Risk often shows itself in the form of carbon emission rates and financial rates. If there is uncertainty in either rate type, the perceived risk goes up, and this may decrease customer investments in the Valley.
- Produce highly competitive, tradable products that are multinational in reach. The bottom line is that the customers' products have to be better, cheaper, higher quality, and more and more sustainable. TVA's partnership and understanding of the customer goal to be more sustainable and competitive is vital in all discussions and decisions.

These common issues that impact sustainable performance provide examples to potential new companies that TVA values sustainable performance within the Valley. This makes TVA more attractive to new companies, resulting in more jobs and investments in the Valley. TVA's efforts have helped recruit or expand the operations of 218 companies in fiscal year 2017. This resulted in attracting and retaining more than 70,000 jobs and $8.3 billion in capital investment to the region [5]. This resulting increase in capital investment for the Valley and load increase for TVA provide evidence that throughout the customer tradeoff process and their resulting sustainable improvement, TVA receives mutual benefit from our customer involvement. TVA is recognized for our willingness to partner with our customers to allow us both to achieve a higher level of sustainable performance. These efforts are a major reason *Site Selection* magazine recognized TVA's economic development team for the 11th consecutive year as a top 10 utility in North America for economic development [19].

Sustainability for TVA and its customers is not just good business; it is *smart business*. TVA recognizes that sustainable business means taking care of the people living in the Tennessee Valley and thinking of the future. Together, TVA, its customers, and its many stakeholders create opportunities in the region that make life better for the people in the Valley.

Author Profiles

Monte Lee Matthews (TVA) Matthews is a high-performance leader with over 20 years of diverse experience at the Tennessee Valley Authority, including leadership positions in engineering, total quality management, and business operations. Matthews is the Deputy Chief Sustainability Officer and currently leads the development and execution of TVA's sustainable business and climate strategies. Prior to joining TVA, Matthews was an architect with CRSS and an energy program manager for DOD-FORSCOM. Matthews has a master's degree in architecture from North Carolina State University, as well as a B.A. degree in Psychology from UNC Chapel Hill. Matthews has also completed Executive Programs at the Kellogg School of Management at Northwestern University, as well as the Wharton School at The University of Pennsylvania. Matthews has been a Board Regent for NAEM and a Steering Committee member for EPRI's Energy Sustainability Interest Group.

Dawn Best (TVA) Best is the industrial account manager (IAM) for TVA's direct-served customers. She works with customers in Mississippi, Alabama, and West Tennessee. Her work focuses on the business relationship with customers, including rates, power contracts, power supply, and TVA programs. She has been with TVA for 31 years. Previous to taking the position of IAM, she served as the Customer Service Manager, Customer Service Engineer, and District Engineer at TVA. Prior to TVA, Best served as the Director of Marketing and Member Services for West Florida Electric Cooperative, and she served as the Industrial Representative for Florida Power Corporation before that. Best holds a bachelor's degree in Electrical Engineering from Mississippi State University.

Laura Doze (TVA) Doze is a program manager for TVA's environmental management systems where she ensures that the company's processes and practices adhere to TVA's environmental standards of excellence. She focuses on process improvement and standardization and also specializes in sustainability and carbon accounting/reporting. Doze has 11 years of diverse experience in the energy and environmental sector working for consulting firms, the governmental sector, and an industry. She double majored in Economics (emphasis environment) and Spanish at

Southwestern University in Georgetown, Texas, earning her Bachelor of Arts degree. She completed her Master of Science degree in Global Energy Management at the University of Colorado, Denver.

Todd Kiefer (Naval Air Station Meridian) Kiefer, USN (ret.) is director of government relations and economic development for East Mississippi Electric Power Association and president of North Lauderdale Water Association. His career in public utilities follows 25 years as an officer and aviator. He has degrees in physics, strategy, and military history; and diverse military experience that spans airborne electronic warfare, nuclear submarines, operational flight test, particle accelerators, Pentagon Joint Staff, war college faculty, and eight deployments to the Middle East and Asia. He spent 22 months on the ground in Iraq and Commanded Al Asad Airbase.

Madhu Ranade (Steel Dynamics, Inc.) Ranade has served as VP and GM in SDI Columbus since September 2014. Ranade joined SDI when it acquired the Columbus mill from Severstal, which he joined in June 2013. Prior to Severstal, Ranade was with ArcelorMittal S.A., where he held several senior leadership positions, most recently, Vice President, ArcelorMittal S.A., and General Manager, Burns Harbor Flat Carbon and Columbus Coatings USA. Previously, Ranade held various management roles at Inland Steel. During his career in integrated steel plants and mini mill, he gained extensive experience in all aspects of the steel business.

Ranade holds a master's degree in materials science and mineral engineering from the University of California, Berkeley, as well as a bachelor's degree in metallurgical engineering from the Indian Institute of Technology, Mumbai.

Karen Utt (TVA) Utt manages the development, guidance, and implementation of agencywide climate change and greenhouse gas policies, strategies, positions, and processes for TVA's Environment and Energy Policy group. She has been with TVA for 6 years. She works closely with senior and TVA executive management to develop and deploy mitigation and adaptation policies and strategies, including the implementation of technical and IT solutions for successful utility operation in a carbon-constrained and climate-resilient business environment. She also coordinates with other TVA business units focusing on climate change legislation, regulations, programs, and initiatives to support TVA's operations. Utt has a Bachelor of Arts degree in biology with a minor in political science from Southwest Minnesota State University and a Juris Doctorate degree from Mitchell Hamline School of Law.

Rob Wise (Resolute) Wise is the General Manager of Resolute Forest Products, located in Grenada, MS. He has been employed by Resolute for 5 years. He earned a Bachelor of Science degree in industrial technology from Mississippi State University and an AAS in Instrumentation from East Mississippi Community College.

Company Profile

The Tennessee Valley Authority is a corporate agency of the United States and the largest public power system in the country. It provides electricity for business customers and local power distributors serving nine million people in parts of seven southeastern states. TVA receives no taxpayer funding, deriving virtually all of its revenues from sales of electricity. In addition to operating and investing its revenues in its electric system, TVA provides flood control, navigation, and land management (80,000 square miles of public lands, 12,000 archeological sites, and 11,000 miles of shoreline) for the Tennessee River system and assists local power companies and state and local governments with economic development and job creation.

References

1. "Tennessee Valley Authority Act," United States Congress, 1933, https://www.tva.com/file_source/TVA/Site%20Content/About%20TVA/TVA_Act.pdf. Accessed 17 December 2017.
2. *Annual Report for the Tennessee Valley Authority for the Fiscal Year Ended June 30, 1934, Ashwander v. Tennessee Valley Authority*, 297 U.S. 288 (1936).
3. "2015 Integrated Resource Plan," Tennessee Valley Authority. https://www.tva.com/Environment/Environmental-Stewardship/Integrated-Resource-Plan. Accessed 17 December 2017.
4. "Sustainability Program," Tennessee Valley Authority Chief Sustainability Officer Interview, for TVA Today publication, March 19, 2017.
5. "TVA Keeps Valley Commerce on the Grow," Tennessee Valley Authority, https://www.tva.gov/Economic-Development/TVA-Economic-Development-Keeps-Valley-Commerce-on-the-Grow. Accessed 17 December 2017.
6. "2016 Fact Book," Tennessee Valley Authority. Section 11, p 2 (PDF page 208). Accessed 17 December 2017.
7. "Carbon Dioxide," Tennessee Valley Authority. https://www.tva.com/Environment/Environmental-Stewardship/Air-Quality/Carbon-Dioxide. Accessed 17 December 2017.
8. "Valley Sustainable Communities," TVA, https://www.tva.gov/Economic-Development/Engage/Valley-Sustainable-Communities. Accessed 17 December 2017.
9. "10-K for the fiscal year ended September 30, 2017, Tennessee Valley Authority, http://www.snl.com/irw/Doc/4063363/Index?did=42643490. Accessed 17 December 2017.
10. "Accelerating Low-Carbon Transition," We Mean Business, https://www.wemeanbusinesscoalition.org/. Accessed 17 December 2017.
11. "Environmental Performance Becomes a Fiduciary Responsibility," *The State of Green Business 2017*, Green Biz, page 31, https://www.greenbiz.com/report/state-green-business-2017. Accessed 17 December 2017.
12. *Out of the starting blocks: Tracking progress on corporate climate action*, CDP, 2016, https://b8f65cb373b1b7b15feb-c70d8ead6ced550b4d987d7c03fcdd1d.ssl.cf3.rackcdn.com/cms/reports/documents/000/001/228/original/CDP_Climate_Change_Report_2016_Web_Pages.pdf?1477406073. Accessed 17 December 2017.
13. "Investing in carbon efficient equities: how the race to slow climate change may affect stock performance," Credit Suisse, 2017, https://www.credit-suisse.com/media/assets/corporate/docs/about-us/responsibility/banking/investing-in-carbon-efficient-equities.pdf. Accessed 17 December 2017.
14. "See transition as an opportunity, rather than a restriction," CDP, 2016, https://www.cdp.net/en/research/global-reports/tracking-climate-progress-2016. Accessed 17 December 2017.

15. "Electric Power Sector Protocol," The Climate Registry, https://www.theclimateregistry.org/tools-resources/reporting-protocols/electric-power-sector-protocol/. Accessed 17 December 2017.
16. "Metals Recycling and Ferrous Resources," Steel Dynamics, Inc., http://www.steeldynamics.com/Sustainability/Environmental.aspx. Accessed 17 December 2017.
17. "Renewable Energy and Sustainability," U.S. Navy, http://navysustainability.dodlive.mil/energy/shore/renewable/. Accessed 17 December 2017.
18. "Green Energy," Resolute Forest Products," http://www.pfresolu.com/Sustainability/Climate_Change_and_Energy/Green_Energy/. Accessed 17 December 2017.
19. A. Brums, "Decision Enhancers: This Year's Top Utilities Back Rich Data with Even Richer Levels of Service and Assistance," *Site Selection Magazine*, September 2017, http://siteselection.com/issues/2017/sep/top-utilities-decision-enhancers.cfm. Accessed 17 December 2017.

Chapter 12
Do We Need Trade-Offs to Advance Climate Policy?

Diane Munns

Abstract Decarbonizing the energy and transportation sectors around the world is not another issue in achieving sustainability; it is *the* issue. Sustainability discussions often pit environmental concerns against social and economic priorities. Balancing climate as an equal concern with economic and societal priorities has achieved incremental change, but not the kind of sea change necessary to address the problem. This chapter suggests that these trade-offs may be a red herring in light of new technology and innovation, largely absent in the electric industry to date. It argues the trade-offs are institutional and behavioral rather than economic. This author makes the case for leadership and provides principles of prioritization and action.

12.1 Introduction

The 2017 update to the National Climate Assessment calls human activity the dominant driver of global warming and warns of a worst-case scenario where seas could rise as high as 8 feet by the year 2100 [1]. The report details climate-related damage across the United States that is already unfolding as a result of an average global temperature increase of 1.8 degrees Fahrenheit since 1900. A second draft document, the *Climate Change Impacts and Risk Analysis*, issued contemporaneously with the National Climate Assessment, finds that high temperatures could lead to the loss per year of "almost 1.9 billion labor hours across the national workforce" by 2090 [2]. That would mean $160 billion annually in lost income to workers. With high levels of warming, coastal property damage in 2090 could total $120 billion annually, and deaths from temperature extremes could reach 9,300 per year or, in monetized terms, $140 billion annually in damage. Additional damage of tens of billions annually could occur to roads, rail lines, and electrical infrastructure, the

D. Munns (✉)
Independent, Des Moines, IA, USA
e-mail: dmunns@edf.org

J. Fox, M. Scott (eds.), *Sustainable Electricity II*,
https://doi.org/10.1007/978-3-319-95696-1_12

report finds. This could all be lessened considerably, the report notes, if warming is held to lower levels.

So it is difficult to discuss trade-offs for climate action as on par with economic and social goals. If the assessment is correct, addressing greenhouse gases is necessary to ensure meeting economic and social goals, not in competition with them. If we continue to compromise on achieving carbon and greenhouse gas reduction goals, we risk exacerbating the climate outcomes that drive severe economic and social impacts.

With all the evidence of human impact on climate, one would assume addressing climate would be approached as an all-hands-on-deck situation. Yet we sometimes appear to be going backward by trying to slow or reverse the positive trend toward decarbonization through revision of environmental rules, subsidization of coal plants, and denial of the human contribution to climate change. The grounding question should always be: What are the consequences of the current political or business view being wrong as opposed to the worst-case scenario advanced in the National Climate Assessment?

If we make the changes necessary to slow and ultimately reverse carbon intensity and we are wrong about this human contribution to climate change, the worst that will have been done is decarbonization of the energy and transportation sectors and a transition to cleaner forms of generation and a more efficient grid. We will have electrified the transportation sector and cleaned up the air. We will have created a more secure, distributed grid that is more resilient. We will have mitigated uneven access to electrification by taking advantage of renewable, fuel-free resources, like sun and wind.

This chapter posits that we are framing the issue on trade-offs incorrectly or at least inadequately. I contend it is not a matter of trade-offs between economics, environment, and social goals that constitute the barrier to progress toward decarbonization necessary to reverse climate trends. Many articles and papers have been written showing that new technology can displace polluting fossil fuels and provide reliable, affordable service [3]. Others have recommended changes in business models from the traditional rate base cost of service model to adequately compensate utilities and accommodate the transition [4]. Many leaders, including those at utilities and major US corporations, have stepped up to say the transition can be accomplished while maintaining affordability and profitability [5].

So what trade-offs are involved that hamper progress? This chapter focuses on institutional considerations, behavioral trade-offs, trading business as usual for change, trading short-term profits for long-term gains, and finally trading known risks for unproven opportunity. These are the trade-offs to be pursued to break the paradigm and embrace the future of clean electrification.

12.2 Perspective for the Chapter

I have been involved in some aspect of the public utility business since 1981. Full disclosure requires my upfront admission that I believe climate change is happening and human activity has a central role. My resume is a bit unusual because I have been both on the staff and a commissioner at a state public utility commission. I have worked as a member of the management team at the Edison Electric Institute, the trade association for the investor-owned utilities, and as a vice president at MidAmerican Energy Company, a Midwestern utility. I have held leadership positions within the National Association of Regulatory Utility Commissioners (NARUC), including president. I now work with the Environmental Defense Fund in its clean energy program. I am trained as a lawyer and function in the policy arena. I am not a scientist, an economist, or an engineer. I am the mother of three children and five grandchildren and fully invested in the future generations who will live with the consequences of our decisions. And I have become the grandmother and AARP member often referred to by consumer advocates as not able to embrace change and needing special protections. My perspective comes from participating in the public arena and debate for over 35 years. I lived through decisions to bring new coal and nuclear plants online and into the rate base, to recover gas take-or-pay contracts, and to restructure the electric industry. I experienced the telecommunication revolution, the rise of energy efficiency and adoption of renewable portfolio standards, the debate over Waxman-Markey and the Clean Power Plan, the appearance of shale gas, distributed generation, and digitization of the electric industry. I experienced these events while serving different roles, although mostly as a state regulator.

Three events had a particularly profound effect on my thinking for this chapter on trade-offs and bear discussion.

12.2.1 New Technology Will Drive Continued Innovation

The evolution of the telecommunication industry from POTS (plain old telephone service) to PANS (pretty amazing new stuff) has transpired at warp speed. It has been since the introduction of the iPhone in 2007 that convergence of modern communication, as we know it, began. I viewed the transition from the regulatory agency charged with overseeing incumbent providers hesitant to change. I argued the first case that went to the US Supreme Court in the late 1990s challenging the federal law that opened up the local telecommunications market to competition. Most striking in retrospect is that it was not the federal law or regulators that led to the modern communications market, but rather digitization of the industry, the Internet, the wireless industry, and the innovation that followed. When a platform was available that made it easy for entrepreneurs to offer new products and services, innovation rapidly followed. This platform was largely outside the grasp of the traditional regulatory structure and regulators. This taught me that we cannot stop or micromanage innovation when customers see products that will improve their lives. New technology will find a way to penetrate the market if it offers a functionality, product, or service desired by the consumer.

Regulation also plays a role: it can slow down or enable the change. Regulators can say no to anti-competitive behavior designed to protect the status quo and discourage new competitors who seek to enter the space. The old Bell System, once powerful and ubiquitous in its domination of the communication sector, offers a cautionary tale. It has been replaced by a myriad of communication and information platforms and convergence of these platforms into mobile, handheld devices that have changed our lives. Ultimately Ma Bell was a victim of denying the change and its own audacity of believing it could stand in its way and deny consumers products they wanted.

12.2.2 Collaboration Can Drive Progress

Energy efficiency (EE) and the negawatt is a concept that has been around for years. Amory Lovins described the potential for efficiency in 1990 in his article, "The Negawatt Revolution" [6]. But many of the programs and much of the infrastructure and progress initially made in efficiency fell away with electric restructuring and the notion that energy efficiency could compete without government programs. In the early 2000s, the efficiency industry had a coastal presence and could be described as a cottage industry with little standardization and high transaction costs. In 2005, Dr. Kathleen Hogan, Division Director at the Environmental Protection Agency (EPA) and later Deputy Assistant Secretary for Energy Efficiency in the Office of Energy Efficiency and Renewable Energy, approached Jim Rogers, CEO at Duke Energy, and me, as President of the National Association of Regulatory Utility Commissioners (NARUC), and asked us to co-chair the National Action Plan for Energy Efficiency (NAPEE) [7]. NAPEE brought together industry and regulators to discuss barriers, incentives, and best practices for advancing energy efficiency. EPA and the Department of Energy (DOE) acted as conveners and, with a relatively small budget, funded meetings and provided technical assistance to drive the dialogue. Since 2008, customer-funded EE program expenditures more than doubled, increasing from $3.4 billion to $7.5 billion in 2016 [8]. While I do not attribute this growth entirely to NAPEE as many factors were in play, neither do I discount the power of combined leadership and collaboration as more and more states, utilities, and consumer advocates created new efficiency programs and goals for efficiency by working through barriers to their deployment.

12.2.3 Aligning Interests Can Advance Change

Significant in my thinking on trade-offs is the phenomenal growth of wind power in Iowa. Iowa initiated the first national renewable portfolio standard in 1983 for a combined state-wide goal of 105 megawatts. The utility companies were initially hostile toward the goal and resisted the mandate through litigation. It was not until there was a combined realization that wind could be profitable for utilities, good for customers, and another source of economic development that the building of wind accelerated. Today, more than 35% of the electric power generated in Iowa now comes from wind power. Iowa had over *6900 megawatts* of wind generation capacity at the end of 2016. In fact, on October 23 and 24, 2017, the wind blew hard enough for wind turbines to provide energy to meet all of MidAmerican's customers' needs in Iowa [9]. The wind industry provides jobs in manufacturing, construction, and maintenance and pays farmers for use of their land. The wind turbine has proven to be compatible with land use and a boon to the state.

From this experience I learned it is better to find ways to align interests than to require companies to take action that is in conflict with their financial interest. Pushing mandates will result in grudging compliance. The requirement will be treated as a regulatory compliance item within the company rather than as a profit center within the business. It is telling that Iowa has never found it necessary to amend its renewable portfolio standard. MidAmerican Energy announced in 2017 that its goal is to eventually produce 100% of its power with renewable resources [10].

With these learnings in mind, the trade-offs to be explored in this chapter are not the usual trade-offs between economic, environmental, and social goals often discussed in sustainability. I believe these arguments are a red herring, misleading or distracting us from the real institutional and behavioral trade-offs impeding progress.

12.3 The Importance of Leadership in a Time of Transition

At the 2016 Edison Electric Institute summit, Detroit Edison CEO Gerry Anderson appeared on a panel to talk about the results of his company's recent carbon planning. As reported in *Utility Dive*, his remarks exemplify the kind of leadership needed to make the transition to a clean energy future [11]. Certain pieces of the reported discussion bear repeating in their entirety.

> I really feel in many ways that our sector would be well served to get out in front of this and let the world know that we've got this one — we will deal with this issue," he said. "And as a result, we really ought to be in a position to pull in the transportation sector's energy demand and general industry's energy demand as well.
>
> We learned there's no sucker's choice here," he said. "You can have a healthy economy and have a healthy environment at the same time.

And he went on to say, “We had to take the first really hard long-term look on how emissions could evolve at our company and what it would cost. What I learned from the process was that we could deeply decarbonize DTE Energy and we could do it in a way that’s affordable. And to be honest, it didn’t take esoteric technology assumptions.”

If a major fossil-dependent utility CEO can publicly make an assertion that it is possible to affordably cut carbon 80% from 2005 levels using known technologies, there appears to be a solid opportunity to avoid trade-offs between profits, customers, and environmental goals. What is needed is the will to change business as usual and the leadership to drive the change and outcome. But it may also mean moving off a known business model that assures short-term profits.

12.4 Support from Powerful Allies

Utility companies will not need to achieve this change alone. Powerful allies are increasingly showing up to support the acceleration of the path forward. The US Climate Alliance, a bipartisan coalition of states and unincorporated self-governing territories in the United States, was formed in response to President Trump’s announcement that the United States would withdraw from the Paris accord [12]. This Alliance has grown to represent 36% of the US population and more than $7 trillion of the country’s gross domestic product [13]. Similarly, America’s Pledge, formed by Michael Bloomberg and California Governor Jerry Brown, demonstrates the intent of state and local governments to continue progress in spite of federal non-action and active hostility toward decarbonization [14].

Some of America’s biggest tech companies renewed their Obama-era commitment to reducing carbon even after President Trump rescinded many of the Obama climate policies. They believe clean energy will not only be good for the environment but also for their bottom line. Apple, Amazon, Google, and Microsoft are some of the 154 companies that agreed to champion clean energy during the 2015 United Nations Climate Change Conference in Paris and who joined the American Business Act on Climate Pledge [15]. These companies do not see a conflict between strong clean energy and climate policies and believe it can be achieved while also supporting American competitiveness, innovation, and job growth.

Investors also see the need for carbon planning and emission reductions. In 2017, a significant percentage of shareholders pushed ten major electric utility companies with large carbon footprints – including PPL Corporation, Dominion Resources, Duke Energy, Southern Company, DTE Energy, and PNM Resources – to undertake analyses to demonstrate alignment between long-term business plans with the goals of the international Paris Climate Agreement and the accelerated transition to a low-carbon economy [16]. These votes are a strong indication that investors understand the material financial risks of climate change and want to understand how the companies in which they invest plan to manage climate risk. So why do we perpetuate the myth of conflict between profitability, affordability, and environmental goals when so many with "skin in the game" believe otherwise?

12.5 Short-Term Gain Versus Long-Term Payoffs

In 2007, the Electric Power Research Institute, led then by Steve Specker, released a discussion paper at its summer seminar [17]. This paper provided stakeholders with a framework for developing a research, development, and demonstration Action Plan that would enable sustainable and substantial electricity sector carbon emission reductions over the coming decades. Based on a "bottoms-up" review of technology performance capabilities and deployment potential, EPRI developed a technical assessment of the feasibility for future US electricity sector carbon emissions reductions, called the PRISM analysis, from the colorful graphic it elicited. The analysis represented an estimate of the potential electric sector emission reductions needed and the technology portfolio necessary to achieve it. This portfolio included:

- End-use energy efficiency
- Renewable energy
- Advanced light water nuclear reactors
- Advanced coal power plants
- CO_2 capture and storage
- Plug-in hybrid electric vehicles
- Distributed energy resources

While these tools remain key to reducing emissions, starkly missing from the PRISM analysis was natural gas. In 2007, it was believed that America's natural gas reserves were in decline. Modern fracking techniques were just being deployed, unlocking a wealth of cheap, domestic natural gas.

With the appearance of inexpensive natural gas, the discussion abruptly changed and natural gas was heralded as the "bridge fuel" to the clean energy economy. Suddenly, rapid advancements in decarbonization were possible, simply by changing out aging coal plants with natural gas plants. But natural gas has roughly half the carbon emission profile of coal. Drilling and transportation releases another powerful greenhouse gas, methane. So natural gas is not a panacea but another tool to be strategically deployed as we move toward deep decarbonization. Natural gas plants have capabilities that can be used to balance the variability of other resources and the availability of the resource should not be discounted in our fight to decarbonize. However, what can and should be questioned is the role it will play going into the future.

Unwise natural gas investments – simply changing out natural gas for coal – can stall the long-term path to decarbonization by locking in natural gas infrastructure and creating an uphill battle for new renewables to compete because of sunk costs. Instead of treating gas as a tool in the toolbox to deep decarbonization, some companies are treating it as the only tool and have abandoned investments in efficiency and renewables beyond what is deemed necessary to meet government standards or to allay critics. They are trading known short-term company profits for long-term stranded investments to the detriment of customers and the environment. Sue Tierney of the Analysis Group warns investment in more long-lived fossil fuel assets could commit the planet to some of the more serious impacts of climate change. "The more we invest in capital stock that is inconsistent with deep decarbonization, the more we are creating incumbents who will fight against addressing decarbonization," she said [18]. Natural gas can become the new coal. Instead of providing a short-term bridge to the future, it could become a bridge too far.

The new trade-offs include evaluating short-term gains with long-term outcomes. Regulators should scrutinize increased investment in long-lived fossil assets with goals for decarbonization and risk to customers. At this time of rapid transition and innovation, companies and regulators should determine if there are alternatives to fossil investments that will leave opportunities open for the future.

12.6 Old Regulatory Paradigms Traded for New Approaches

Regulators are often the most resistant to change. Most prefer to be comfortably in the pack rather than out front. To a certain degree it is understandable, as regulators are not celebrated for innovation that works but are blamed for change that goes astray. So often the first question in a regulatory proceeding is, "What have other states done and what was the outcome?" But it is often inexplicably difficult to get regulators to make decisions that should not be difficult to make.

A case in point is voltage optimization. There is technology currently available to companies that can more precisely monitor and manage required voltage levels, the result being the ability to lower voltage with a corresponding drop in customer usage and customer bills. The voltage drop is undetected by customers and made possible by new, cost-effective technologies that can more precisely measure and adjust voltage. Instead of providing higher voltage to ensure adequate support, support can now be provided with more precision. Reports of estimated savings are 3–5% [19]. Compare that to a good efficiency portfolio that will achieve a 1–2% savings. So why hasn't this technology been deployed and these savings captured for customers and the environment?

It is because of relic incentives that compensate utilities based on usage. The more energy customers use, the more money the company makes. It may be understandable why companies do not advance voltage optimization as it cuts usage and sales. But that also begs the question of leadership and changing the status quo. This technology should be universally advocated by consumer advocates and required to be deployed by regulators. Companies are required to provide safe and adequate service under the regulatory compact. Shouldn't prudency in meeting that standard impose a duty on companies to bring forward cost-effective technology that will optimize the system, even if it cuts profits? Companies are more than willing to invest in equipment that leads to higher customer bills and argue for recovery based on operational need. Should the converse not also hold that when there are technologies that increase operational efficiency? Companies should be held to a standard of prudency to evaluate and deploy these cost-saving technologies. Regulators and consumer advocates must be willing to advocate for these new technologies and require companies to invest in them, without an additional incentive beyond a return on the cost of the equipment. Ideally, companies would bring this solution forward as a cost-effective way to meet emission reduction goals. This is an example of changing the status quo to embrace new technologies and opportunities.

12.7 Risk-Reward Assessment

Risk or loss aversion is a preference for a sure outcome over a gamble with higher or equal expected value. Some studies have shown people typically exhibit greater sensitivity to the possibility of losses than to the opportunity for equivalent gains when making decisions [20]. So how do these tendencies play out with regulated utilities and particularly the regulatory and consumer advocacy segment during a transition? In my experience, risk aversion proves a major trade-off in regulation because of the possibility of failure. There is significant reticence to change, much of it attributable to long memories of political backlash when attempted change did not accomplish what it set out to do. So we are rewarded when we do nothing and do not risk the chance that something will go wrong. But in doing so, we also forego the opportunity to do something right. As studies explain, losses can be twice as powerful psychologically as gains. With change upon us, we need to collaborate and move forward together to use new technologies and drive innovation to meet goals.

A good example of risk aversion impeding needed progress is with the slow adoption of time-differentiated or time-of-use pricing. The cost and carbon profile of dispatch of generation resources changes with the usage of the customer base, but pricing does not reflect this change. The concept of electricity providers sending customers price signals to change usage behavior to help manage costs and carbon has been around for decades. It has been the subject of any number of pilots demonstrating value through the years but remains a seldom-used option. The importance of utilizing this tool becomes more significant as we anticipate increasing usage through electric transportation. Successful time-of-use pricing has become more possible with the information available through the millions of smart meters deployed throughout the country. The information from these meters makes two-way communication possible and also makes it possible to understand the impacts of pricing changes at a granular level.

But there is extreme angst that leads to sustaining flat pricing and rejecting mandatory time-differentiated pricing, even when the significant benefits that could be achieved for customers and the environment are understood. If a study shows 90% of customers will pay less on time-of-use rates, even if they do nothing, some will oppose the change by focusing on the 10%. We need to stop letting perfection be the enemy of the good. This is an institutional trade-off that needs to take place as we transition, understanding there may be missteps. It doesn't mean we ignore the impact on the 10%, but that we understand who they are and what can be done to mitigate any negative impacts. Fear of a misstep leaves us paralyzed, standing in place.

12.8 Conclusion

In a time of change, the first reactions to ideas that threaten the status quo are to circle the wagons and protect. This was the reaction seen in the telecommunication industry when new players and businesses wanted to interconnect with the legacy system. Companies employed their knowledge of the regulators and the regulatory system to make it difficult for new service providers to engage. Instead of thinking about how their system could be used as a platform and viewing new entrants as potential partners or customers, the impulse was to squash the competition to maintain control. That attitude drove communication off the landline system and stranded once profitable equipment. As we move to welcome new players onto the electric system to help in achieving the climate solution, here are some tips to use in trading off behaviors and attitudes that protect the status quo for those that can assist in making the necessary transition to a sustainable, clean energy economy.

- Establish long-term climate goals and test all decisions for consistency with the path.
- Question the applicability of underlying assumptions in the new paradigm.
- Reject locking in today's resources because that narrows rather than widens future opportunities and choices.
- Require cost-effective measures like energy efficiency and voltage optimization to be widely deployed.
- Take shorter-term actions to manage risk. Ask if there is an alternative to business as usual that will either delay committing to a long-term fossil decision, offer multiple functionalities for achieving future action, or just be a better solution.
- Open companies and regulatory proceedings to new players and consideration of new technologies.
- Collaborate whenever possible and look for opportunities to turn potential competitors into potential partners.

We are coming to realize we have the means to achieve our climate goals and avoid the worst impacts of climate change. But we are running out of time. I have performed many of the roles discussed in this chapter and understand they are difficult jobs. Now we need the will, the courage and the strength to make the change.

Author Profile

Diane Munns Munns has been involved in some aspect of utility regulation for the past 35 years. She began her career at the Iowa Utilities Board in its general counsel's office. She became General Counsel to the Board and eventually was appointed to the Board and served as its chair. While on the Board, she served the longest term in history as president of the National Association of Regulatory Utility Commissioners. Following her 20-year tenure at the Iowa Utilities Board, she worked at the Edison Electric Institute and MidAmerican Energy Company and now is employed by the Environmental Defense Fund as its Senior Director for External Affairs and Regulatory Policy in the clean energy program. Key career milestones include arguing before the US Supreme Court on behalf of state commissions; developing the first regional state committee, the Organization of MISO States; co-chairing the National Action Plan for Energy Efficiency; and establishing the Edison Foundation's Institute for Electric Efficiency, now known as the Institute for Electric Innovation. Munns has held numerous leadership positions and has received awards for her work, including the United States Energy Association's Public Service Energy Leadership Award and the Midwest Energy Efficiency Alliance's Inspiring Efficiency Leadership Award. Munns lives in Des Moines, Iowa, and provides public service as a member of the Des Moines Water Works Board of Trustees. She received her B.A. from the University of Iowa and J.D. from Drake University.

References

1. Wuebbles, D.J., D.W. Fahey, K.A. Hibbard, B. DeAngelo, S. Doherty, K. Kayhoe, R. Horton, J.P. Kossin, P.C. Taylor, A.M. Waple, and C.P. Weaver, 2017: Executive summary, In: *Climate Science Special Report: Fourth National Climate Assessment*, Volume I [Wuebbles, D.J., D.W. Fahey, K. A. Hibbard, D.J. Dokken, B.C. Stewart and T.K. Maycock (eds.)]. U.S. Global Change Research Program, Washington, D.C., pp. 12–34, doi:https://doi.org/10.7930/JODJ5CTG, https://science2017.globalchange.gov/. Accessed 14 February 2018.
2. *Multi-Model Framework for Quantitative Sectoral Impacts Analysis: A Technical Report for the Fourth National Climate Assessment*, United States Environmental Protection Agency, EPA 430-R-17-001, May 2017, https://cfpub.epa.gov/si/si_public_file_download.cfm?p_download_id=532962. Accessed 14 February 2018.
3. Kaufman, Leslie, "Is 100% Renewable Energy Feasible? New Paper Argues for a Different Target," Inside Climate News, 19 June 2017, https://insideclimatenews.org/news/19062017/100-percent-renewable-energy-climate-change-targets. Accessed 15 February 2018.
4. "Ratemaking and Utility Business Models," America's Power Plan, http://americaspowerplan.com/power-transformation-solutions/ratemaking-and-utility-business-models/. Accessed 14 February 2018.
5. Pyper, Julia, "Large Corporations Are Driving America's Renewable Energy Boom. And They're Just Getting Started," Greentech Media, 10 January, 2017, https://www.greentechmedia.com/articles/read/large-corporations-are-driving-americas-renewable-energy-boom#gs.8NYf0OA. Accessed 15 February 2018.

6. Lovins, Amory B. "The Negawatt Revolution," *Across The Board, The Conference Board Magazine* , Vol. XXVII No. 9, September 1990, https://www.rmi.org/wp-content/uploads/2017/06/RMI_Negawatt_Revolution_1990.pdf. Accessed 14 February 2018.
7. "National Energy Plan for Energy Efficiency," United States Environmental Protection Agency, https://www.epa.gov/energy/national-action-plan-energy-efficiency. Accessed 14 February 2018.
8. *Energy Efficiency Trends in the Electric Power Industry*, The Edison Foundation, Institute for Electric Innovation, December 2017, http://www.edisonfoundation.net/iei/publications/Documents/IEI_Energy%20Efficiency%20Report_Dec2017.pdf. Accessed 14 February 2018.
9. "High Winds Bring Milestone for Iowa Utility's Wind Power," U.S. News and World Report, 28 October 2017. https://www.usnews.com/news/best-states/iowa/articles/2017-10-28/high-winds-bring-milestone-for-iowa-utilitys-wind-power. Accessed 14 February 2018.
10. Chow, Lorraine, "Iowa's Largest Utility Eyes 100% Renewable Energy Goal," EcoWatch, 2 May 2017, https://www.ecowatch.com/iowas-largest-utility-eyes-100-renewables-goal-2387877147.html?utm_campaign=RebelMouse&utm_medium=social&utm_source=facebook&utm_content=EcoWatch. Accessed 14 February 2018.
11. Bade, Gavin, "EEI 2017: The utility sector's business case for deep decarbonization," Utility Dive, 13 June 2017, https://www.utilitydive.com/news/eei-2017-the-utility-sectors-business-case-for-deep-decarbonization/444873/. Accessed 14 February 2018.
12. "2018 Strategy," United States Climate Alliance, https://www.usclimatealliance.org/. Accessed 14 February 2018.
13. 2017 Annual Report: Alliance States Take the Lead, United States Climate Alliance, https://static1.squarespace.com/static/5936b0bde4fcb5371d7ebe4c/t/59bc4959bebafb2c44067922/1505511771219/USCA_Climate_Report-V2A-Online-RGB.PDF. Accessed 14 February 2018.
14. "America's Pledge," https://www.americaspledgeonclimate.com/. Accessed 14 February 2018.
15. White House Announces Additional Commitments to The American Business Act on Climate Pledge," White House Press Release, 1 December, 2015, https://obamawhitehouse.archives.gov/the-press-office/2015/12/01/white-house-announces-additional-commitments-american-business-act. Accessed 14 February 2018.
16. Bakal, Dan, "The Power Sector Must Heed Shareholder Calls for 2-degree Scenario Analysis," Utility Dive, 27 June 2017, https://www.utilitydive.com/news/the-power-sector-must-heed-shareholder-calls-for-2-degree-scenario-analysis/445950/. Accessed 14 February 2018.
17. "The Power to Reduce CO_2 Emissions, The Full Portfolio," Discussion Paper Prepared for the EPRI 2007 Summer Seminar, Electric Power Research Institute, August 2007, http://mydocs.epri.com/docs/public/DiscussionPaper2007.pdf. Accessed 14 February 2018
18. Bade, Gavin, "EEI 2017: The utility sector's business case for deep decarbonization," Utility Dive, 13 June 2017, https://www.utilitydive.com/news/eei-2017-the-utility-sectors-business-case-for-deep-decarbonization/444873/. Accessed 14 February 2018.
19. *Evaluation of Conservation Voltage Reduction on a National Level*, Pacific Northwest National Laboratory, PNNL-19596, July 2010, https://www.pnnl.gov/main/publications/external/technical_reports/PNNL-19596.pdf. Accessed 14 February 2018.
20. Kahneman, D. and A. Tversky, A "Choices, Values and Frames," American Psychologist. 39 (4): 341–350, 1984, http://web.missouri.edu/~segerti/capstone/choicesvalues.pdf. Accessed 14 February 2018.

Index

J. Fox, M. Scott (eds.), *Sustainable Electricity II*,
https://doi.org/10.1007/978-3-319-95696-1

D

E

T

MIX
Papier aus verantwortungsvollen Quellen
Paper from responsible sources
FSC® C105338

If you have any concerns about our products,
you can contact us on
ProductSafety@springernature.com

In case Publisher is established outside the EU,
the EU authorized representative is:
Springer Nature Customer Service Center GmbH
Europaplatz 3, 69115 Heidelberg, Germany

Printed by Libri Plureos GmbH
in Hamburg, Germany